I0476254

Digestive System

What is the Digestive System?

The digestive system is a set of organs; which take part in digesting our food and giving us the necessary nutrients.

What is the role of the Digestive System?

The role of digestion is to break down large molecules of food into its simplest small molecules; so that its nutrients can be absorbed; which helps the whole body to thrive and function.

Organs that create this System

- Mouth
- Oesophagus (Gullet)
- Stomach
- Gall Bladder
- Bile Duct
- Liver
- Pancreas
- Duodenum Ileum (Small Intestine
- Colon (Large Intestine)
- Rectum

Types of Digestion

There are 2 main types of digestion; and these two are called Mechanical and Chemical.

1. **Mechanical:** Involves the mouth, teeth, and oesophagus.

2. **Chemical:** Involves the saliva in the mouth, stomach, stomach acid, small intestine, colon, pancreas and enzymes.

Digestive System Procedure

Mechanical Digestion

1. Ingestion

Ingestion is a process; where food is entered though the mouth, settles in the buccal cavity, and chewed up into smaller pieces by the teeth. In the process of chewing, the walls of the mouth, which are the cheeks, contain ducts which secrete a clear digestive juice called saliva. This saliva is secreted by three pairs of glands in the mouth called salivary glands; which have three separate names, therefore, these three pairs are called parotid, sublingual, and submandibular salary glands. This saliva contains amylase enzymes (ptyalin); whose main role is to break down the carbohydrates and starch in foods. This break down of food increases the surface area for enzymes to work on, also, the tongue plays a part in breaking down food; since they are able to taste the food and spread even more digestive juices over the food.

1. Swallowing

When the food has been broken down by the saliva; the swallowing process starts. It all begins when the tongue presses upwards and backwards against the roof of the mouth, therefore, this turns the mechanically broken down food into a ball called a bolus. This bolus will then go to the back of the mouth and go down the oesophagus, therefore, when it starts to go down the oesophagus; a soft palette in the nasal cavity closes itself, the rings of cartilage in the larynx pulls itself upwards, the opening of the windpipe (trachea) will lie itself under the back of the tongue, the epiglottis of the windpipe will partly close; due to the contraction of ring cartilages, and finally the epiglottis helps the food to go down into the oesophagus. The epiglottis is important, since it stops food from going down the trachea, and causing the individual to choke.

Peristalsis

Peristalsis is a process where the oesophagus has layers of muscle; which make sets of contraction for the bolus of food to go down. There are two types of muscle located in the oesophagus; one is the circular muscles; which circulates the canal; while the other is the longitudinal muscles; which runs along the oesophagus. This process takes about six seconds to complete, before the food goes into the stomach, which therefore, ends the mechanical digestion and starts the chemical digestion.

Mouth Enzymes

Name	Location	Function
• Ptyalin	Mouth	Breaks starch to simple sugars.
• Amylase	Mouth	Breaks start and carbohydrates into simple sugars.
• Betaine	Mouth	Keeps cell fluid balances in perfect condition.
• Bromelain	Mouth	Anti-Inflammatory agent.

Chemical Digestion
2. Stomach

Once the bolus of food has gone though the 6 second process of Peristalsis, it finally reaches our stomach. The stomach has elastic walls which will stretch; as the bolus of food comes though and enters the gastric acids in the stomach. These gastric juices are secreted from glands in the lining of the stomach; also, a diluted solution of hydrochloric acid is also secreted by the glands of the stomach lining.

How does this Secretion Happen?

This secretion of gastric juices, and a weakened version of hydrochloric acid; comes from the impulses of the brain. When an individual smells, sees, or tastes food; impulses from the eyes, nose and mouth are sent up to the brain, therefore, once the brain receives these impulses, it will send new impulses back down to the stomach, where it tells the stomach to start secreting gastric juices. Once the bolus food reaches the stomach, a hormone called gastrin is produced and circulates in the blood stream; therefore, its role is to keep stimulating the gastric glands so that they secrete more and more of the gastric juice.

Stomach Acid: As we eat food, we ingest many different types of bacteria which can cause illnesses and affect the body, fortunately, most of these bacteria are killed by the stomachs acid. Our stomach lining produces hydrochloric acid which makes a weak gastric juice; therefore, the role of this acid is to kill many of the bacteria that are on our food. However, there's another question we may ask on this subject, and that is why hasn't our stomach been digested by the hydrochloric acid since this acid is capable of burning and digesting some metals!

The reason our stomachs are not digested by the acid, is because of the body's marvellous way of keeping everything safe. This means our stomach wall has a lining containing epithelial cells that secrete protective substances of mucus and bicarbonate on top of the first layer of the stomach wall, which is protein. The mucus provides a physically barrier, so the acid can't touch the protein wall itself, while the bicarbonate, which is highly alkaline and is secreted by the stomach cells, neutralizes the acid.

Prevention of Self-Digestion

The proteins which create protease in the stomach are not digested therefore, why is this so? The reason for this, is because of the proteins marvellous way of making the protease in a inactive form, therefore, this inactive version of protease is called pepsinogen; which does not become active until its enters the gastric juices and mixes with the weakened hydrochloric acid. The mucus on the stomach lining also protects the whole stomach from the active pepsinogen.

Chemical Digestion of the Stomach Continued

Once the bolus is in the gastric juices of the stomach; the muscle of the stomach goes though peristaltic movements; to churn and mix up the food into a creamy juice. The time it takes for the food to be churned up in there, depends on each individual stomach, therefore, once the food is creamy; the juice will pass though the pyloric sphincter a little at a time to the small intestine (Duodenum). Other roles of the pyloric sphincter, is to stop large molecules of food going through to the small intestine.

3. Duodenum/ Small Intestine

When this creamy food flows into the small intestine, the brain will do the similar secretion technique with the walls of the small intestine, however, the hormone secretin will be made to flow in the blood stream, be stimulated by the acid contents of the stomach, and stimulate the walls of the Duodenum to create pancreatic juices. The pancreas lies below the stomach, and it creates the many different types of enzymes; to break down the different types of food into smaller molecules.

Intestine Enzymes:

Enzyme	Location	Function
Protease	Intestine	Breaks down protein to amino-acids.
Lipase	Intestine	Breaks down fat to simple lipids
Amylase	Intestine/Gut	Breaks down carbohydrates to simple sugars.

Other Roles of the Pancreas

Other roles of the pancreases are:

- **Neutralize Gastric Juices in the stomach:** Pancreatic juices are made from Sodium Hydrocarbonate; therefore, this juice is able to partly neutralize the gastric juices in the stomach.

- **Adjust the pH for Pancreas Enzymes:** What this does is give the pancreas enzymes the right temperature to thrive in; since a high ph would cause then to denature.

4. Small Intestine: Absorption

When the creamy food is broken down into smaller molecules; little hairs which have microvilli, lacteal, and are on the walls of the small intestine; will start to take up the simple molecules. Therefore, these hairs are called villi; and the process where the simplest molecules are sucked up for nutrition is called absorption. In absorption, the walls of the small intestine are large absorbing surfaces, with circular folds in its internal surface. Once the molecules are sucked up by the villi; the molecules are then taken into epithelial cells; before being taken though the wall of the capillaries in the villi and into the bloodstream. Once the molecules are in the bloodstream, they are finally carried away to a network of veins called the Hepatic Portal Vein. What this portal vein does is carry the molecules in the blood to the liver; where it shall be released into the general bloodstream.

Liver
The liver lies just underneath the diaphragm of the lungs and overlaps the stomach. Such roles of the liver are:

- **Regulates sugar in the Blood:** If too much sugar is in the blood stream, the livers role is to take the excess sugar and store it in the body as glycogen. If the blood is low on sugar, the liver will take the glycogen, covert it to glucose, and send it back to the bloodstream; therefore, this makes the glucose level of the blood steady again.

- **Create Bile:** Liver cells are in change of making bile in the liver, before sending it to the gall bladder; to be released into the small intestine though small bile ducts. Bile is also released to assist digestion of fats.

- **Store Iron:** The haemoglobin from dyeing red blood cells is stored in the liver.

- **Create Plasma Proteins:** Most of plasma in the body; such as Fibrinogen in blood clotting, is made from the liver.

- **Dextox the Body:** Compounds or substances that are poisonous are transformed into harmless substances by the liver. The poisonous compounds can rather come from bacteria; which have acted on amino-acids, or from environmental foods such as drugs. Once this detoxification is done, it's sent to the kidneys to be excreted via the renal system.

- **Store Vitamins:** Vitamins A and D are located in the liver. This is also the reason why the liver of an animal is the best organ to get vitamin and nutrients from.

5. Small Intestine: Assimilation
Assimilation is where the molecules of different types of food are taken to be used in the body.

Types of food that are taken into Assimilation
1. **Minerals:** Calcium, Magnesium, Chloride, Potassium and Iron are important because they are needed for our body functions, for example, the functions such as strengthening our bones and teeth, creating muscle, creating nerves and the constructions of red blood cells.

2. **Carbohydrates:** Fruits, vegetables, breads, potatoes and cereals contain important carbohydrates that make up 70% of the mass in a human diet and are what humans need in order to live, are active and have energy. These large organic compounds are made from the elements carbon, hydrogen and oxygen. It's not only humans that need carbohydrates; plants need carbohydrates too because in the process called photosynthesis, the plant stores these carbohydrates as starch into their stems. Carbohydrates, which are broken down into glucose, are used for cell respiration, contraction of muscles and electrical changes in the nerves.

3. **Proteins:** Protein foods such as meat, fish and Soya make up 16% of our body weight and are all good for the growth and repair of the human body, cells, hair, skin and connective tissue. Proteins are similar to carbohydrates, except that the organic compound structure of protein contains nitrogen, followed by carbon, hydrogen and oxygen. Proteins will be

used to create plasma in the blood; as well as create cell structures, be fuel for energy, and the cell membrane.

4. **Lipids/Fats:** Saturated and un-saturated foods such as sweets, oils, junk food, omega 3 and 6, are important for the body's insulation and protecting the body's vital organs. Without fat, humans would be affected more in cold weather as well as suffer Anorexia Nervosa or Bulimia. Another use of lipids, are to create the membranes and structures of cells; as well as assist in cell metabolism.

5. **Vitamins:** Vitamins are very important for the body because they support the body's main functions, for example, vitamin A, B, C and D helps humans to see in dim light, release energy from food in respiration, keep bones, skin and teen healthy as well as help absorption of calcium. Vitamins are not made in our bodies, so it vital that it is included in an overall healthy diet.

6. **Excretion**

Large Intestine
Once all the small molecules have gone though absorption and assimilation, the remaining food, which is mainly roughage, fibre, vegetable fibres, mucus, dead cells, and water, goes into the large intestine. The reason why cellulose cannot be broken down and digested is because our body has not created the right type of enzymes that break down cellulose and roughage. The bile salts and emulsified fats are reabsorbed by the ileum.

7. **Excretion: Rectum**
Finally, what's left of the food is a solid called faces, therefore, its passed though the rectum by peristalsis; before it goes though the process of defecation; in which the faces is pushed out of the anus.

Respiratory System
What is the Respiratory System?
The respiratory system; is the system that helps us to breathe. The main organ of this system, the lungs, is located inside of the thorax and is covers by the ribcage and muscular sheets.

What is the Role of the Respiratory System?
The role of the respiratory system is to take oxygen, for cellular respiration; while taking out wastes such as carbon dioxide. Another role of this system is to keep bring oxygen in, so that cell respiration takes place to give nutrients and energy the other organ cells and muscles cells of the body.

Organs that create this System
- **Nose/Mouth**
- **Trachea**
- **Ring of cartilage**
- **Ribcage**
- **Lungs/Ribcage and Diaphragm**
- **Bronchus**
- **Bronchioles**
- **Alveoli's**

System Procedure

1: The air we breathe in first flows through the nose or mouth. Once the air is breathed in; it will go down the trachea and into the lungs.

Characteristics of the Nose and Trachea

The table below shows the characteristics of the nose trachea.

Table:

Characteristic	Information
Rings of Cartilage	These rings are c shaped; and are located around outside of the trachea and bronchi; therefore, if you imagine the outside c shaped rings on a hover, they are just like that. Their role is to stop the trachea from collapsing on itself; therefore, it holds the structure of the trachea.
Cilia	These are small hair like structures located in the nose. Their role is to sweep trapped dirt and particles away from the trachea and towards the mouth. This stops any microorganism and pathogens from going into the airways of the lungs and causing infection.
Epithelium	A lining inside of the trachea, bronchi and bronchioles which secretes mucus. The mucus is secreting from ciliated cells and this mucus form a layer over the internal lining to stop bacteria and other microorganisms from causing infections. Once they trap the microorganisms, the mucus with them in it is carried back up the lungs by the cilia before reaching the top of the trachea and going down the gullet to be digested.

2: Once the air flows from the trachea into the lungs; the air will separate down two airways called the bronchi, therefore, each bronchi leads to the one side of each lung. The air will continue to flow though the bronchi and separate down many smaller airways called the bronchioles. Finally, the air will reach the alveoli where the exchange of oxygen and carbon dioxide takes place.

How does our Ribcage and Lung work in Ventilation?

In ventilation; there are two movements which the ribcage does to take in oxygen; and breathe out carbon dioxide. These two movements are called inhalation and exhalation; also, the muscles that help create the movements are called the intercostals muscles and the external intercostals muscles. These muscles are very important, since their elasticity is what helps take more air in; and breathe more carbon dioxide out of the body.

Inhalation

When a person inhales, they are starting to take in oxygen from their atmosphere, all the way down to the alveoli's. Therefore, what happens in this process; is that the ribs will move upwards and outwards. This movement happens by the external intercostals muscles contracting, and pulling the ribcage up. As the ribcage pulls up, the volume of air inside of the lungs is increased; therefore, this causes the diaphragm of the lungs to pull down into a flat shape; so that the air can fully enter the lungs. In an adult, they can breathe in up to 5 litres of air from their atmosphere.

Types of Pleural in the Lung

In our lungs, there are many types of pleural membranes and fluids which keep our lungs and ribcage safe from damage. The table below explains these types in more detail.

Pleural Type	Information
Pleural Membrane	A membrane forming a protective envelope around the lungs.
Pleural Fluid	A lubricant to stop the lungs and ribcage from sticking to the inside of the chest.
Pleural cavity	The space between the membrane and fluid/

3: Change: Alveoli's

There are over 700,000,000 of these alveoli's in each lung. The alveoli's are small air sacs which bunched together like grapes, and are responsible for taking oxygen to the red blood cells; while ejecting carbon dioxide into the plasma; so it will be breathed back out into the atmosphere. When the oxygen hits the alveoli and is diffused to the red blood cells; the oxygen will mix with the haemoglobin of the red blood cells and forms oxyheamoglobin, therefore, this oxygen will be taken though the capillaries, to the pulmonary vein, before going to the right atrium of the heart. The alveoli's also have a film of moisture which lets the oxygen diffuse into the red blood cells. Not all oxygen is removed from the air, on average; the human will breathe in 21% of air but only will emit 16% of the oxygen back into the atmosphere.

4: Exhalation

Carbon dioxide, created from the after product of respiring cells and tissues, is ejected from the blood plasma into the alveoli's. This is where the individual will breathe out, send the carbon dioxide all the way the alveoli's, bronchioles, bronchus, trachea; before finally out of the mouth. Other type of gases, which is sent to the alveoli's in exhalation; are nitrogen, water vapour, oxygen and argon.

How does our Ribcage and Lung work in Ventilation?
Exhalation

In the process of exhalation, the opposite happens. Therefore, when an individual breathes out carbon dioxide; the internal intercostals muscles of the ribcage contract, while the external intercostals muscles relax. Since carbon dioxide need to be pushed out from the lungs, the diaphragm will relax and recoil back into it dome shape. At the same time as the diaphragm does this, the volume of air in the lungs decreases.

Characteristics of the Respiratory System

Characteristic	Information
Large Surface Area	Since there are a total of 700,000,000 alveoli's on each lung, the surface area would come to 60 squared. If we where t compares this surface area in real life; the surface area would 60 square meters which is larger than a classroom.
Ventilation	Ventilation of the lungs and ribcage helps keep the body functioning and living.
Capillary Network	The capillaries, which the oxygen and carbon dioxide is carried, helps in getting rid of waste and supporting cell respiration

Summary of the Body Systems

Body System	Role

Digestive System	Break down food for nutrients; therefore, keeps the body active ad in good condition.
Circulatory System	Pumps blood around the body. The red b carries oxygen in and carbon dioxide out in to keep us alive.
Renal System	Excretes fluid waste and toxins from our bo so we don't die from internal poisoning.
Respiratory System	Keeps our muscles in good condition by ta out carbon dioxide and brining in oxygen. also helps with cellular respiration.

How does the Respiratory and Cardiovascular/Circulatory System Work Together to maintain Bodily Functions?

The way the Respiratory and Cardiovascular/Circulatory System work together to maintain the body's functions is though cell respiration. Not only can humans do this, but also animals and plants can do this. The differences between an animal and a plant in respiration, is that plants gain their chemical energy though sunlight and the process called photosynthesis.

Cell Respiration
What is Cell Respiration?

Cell respiration is a process, where the circulatory and respiratory systems work together to give cells their needed chemical energy. Therefore, this helps the cells to carry on doing their bodily functions, such as creating new tissue; carrying oxygen to the muscles of the body they also remove carbon dioxide and other wastes out from the body.

Types of Cell Respiration

- **Aerobic Respiration:** This type of respiration requires oxygen to give the cells energy. In this type of respiration, three processes take place; therefore, these three processes are Glycolsis, Kerbs cycle and Electron Transport Phosphorylation.
- **Anaerobic Respiration:** This type of respiration does not require any oxygen to give the cells energy. The cells located in the organs, tissue, and muscles; will respire themselves in order to create energy to carry out their bodily functions. This normally happens when the individual is doing high levels of cardio or exercise.

How does Cell Respiration Work?

The chemical energy our cells need comes from the food we eat. Therefore, in a way our digestive system is also a part of the cell respiration process; which gives our cells the needed energy to carry out the many bodily system functions. When an individual eats a piece of food, it goes down the oesophagus, stomach and the other digestive organs of the body. Therefore, as the food is broken down by certain enzymes in the body; oxygen; which has been breathed in from the respiratory lungs oxidises the now broken down food. In the process of food oxidisation, most of the food is broken down into glucose; therefore, this is where the chemical energy, needed for the cells, is stored. This glucose will be converted into a chemical molecule called Adenosine Tri-sulphate (ATP), therefore, once the glucose has been converted to ATP; it's absorbed by the cells, which gives it energy to carry on its bodily functions.

The Three processes in Aerobic Respiration
- **Glycolsis**

This is a more detailed process of how the glucose is oxidised down, and turned into ATP to gives cells energy in aerobic respiration. When glucose is broken down, three carbon molecules called pryuvate are made; followed by four ATP molecules, two more ATP molecules and two molecules called NADH. The NADH is actually the molecules which carry two energy rich electrons needed by the cells to have energy.

- **Kerbs Cycle**

In Kerbs cycle; the two NADH molecules are put though a complicated set of reactions to produce eight more NADH molecules, therefore, meaning there will be sixteen more energy rich electrons for the cells. From this process, two molecules called FADH2 will be made as well.

- **Electron Transport Phosphorylation.**

In this process, the extra NDAH and FADH2 molecules will travel to the cells special membranes, where molecules (called the electron transport system) will take the energy rich electrons and use them to make more ATP.

Process in Anaerobic Respiration

Because anaerobic respiration means no oxygen is needed to do it, the only process that takes place, is the Glycolsis Process. Another small difference is that only two ATP molecules will be made from this type of respiration.

Diffusion and Active Transport
What is Diffusion?

Diffusion is a process in respiration, where material, such as oxygen, water, nutrients, waste and carbon dioxide, move though the cells membrane to get to other types of cells. Therefore, this movement of the materials in and out of the cell depends on the concentration gradient of the cells membrane. If the concentration of the membrane cells is high; it causes the material in the cell to diffuse out of it.

Active Transport

For cells that want to take up the materials, for example, organ cells and tissues cells, they use a process called Active transport this is where the cell uses energy from respiration to take in the needed materials.

Equation for Respiration

Glucose + Oxygen + Carbon Dioxide + Water (+energy)

Symbol Equation for Cellular Respiration

C6H12O6   6O2   6CO2   6H2O (+energy)

Breathing Rates

Breathing State	Breathing times per minute
Resting	16
Exercise	20-30

How does the Circulatory System help to maintain the Bodies Functions?

The circulatory system helps maintain the body's functions by carrying the red blood cells all over the body to gives nutrients, oxygen, as well as taking in carbon dioxide. It then goes back to the lungs; where the respiratory system takes place and sends the respiratory waste out of the body. When the circulatory system takes oxygenated blood to the muscles and organs of the body, it's helping the cells give nutrition to the muscles and organs to keep them contracting. This produces movement; builds large proteins, for example, enzymes; and supports the process of cell division.

Contraction and Movement of the Muscles

This is all done by our red blood cells providing the muscle cells with oxygen and nutrients the muscles need. If there is a lack of oxygen in the muscles; the muscles cells can do anaerobic respiration on their own. However, overworked muscles, which stop oxygen reaching the muscles to support them; can cause glucose to be converted to lactic acid.

Lactic Acid

Lactic acid is an acid which sends the signal that you are overworking too hard and need to slow down, therefore, this lactic acid will build in the blood stream and cause the individual to have cramps. The cramps disappear by the person resting. Therefore, this lets the red blood cells give oxygen to the muscles cells as well as fully oxidise the lactic acid, and decrease the oxygen debt that is needed in the muscles. In anaerobic respiration; the muscles will produce large amounts of carbon dioxide followed by other wastes and water. Therefore, the cells take the waste from the muscles, before going back to the lungs.

Circulatory System
What is the Circulatory System?

The circulatory system involves our heart, lungs, arteries, vena cava, aorta, veins, capillaries and red blood cells.

What is the role of the Circulatory System?

The role of the circulatory system is to keep us alive by pumping blood around the body. Without any blood; we would be unable to receive oxygen and nutrients, which are carried by our red blood cells to the muscles; therefore, we would not be alive; due to our blood cells not carrying any oxygen to support our muscles and heart.

Organs that create this System

- **Heart:** Main pump; which pumps oxygenated and deoxygenated blood to the lungs and around the body.

- **Lung:** Provides red blood cells with oxygen.

- **Aorta:** Main artery to carry oxygenated blood to the arteries.

- **Vena Cava:** Main vein to carry deoxygenated blood.

- **Veins:** Carry blood from the heart to the lungs. It also carried blood back from the lungs to the heart again.

- **Arteries:** Carries blood to the muscles, limbs and organs of the whole body. They also carry blood back to the heart.

- **Coronary Arteries:** They are connected all over the heart; to let red blood cells give oxygen and nutrients to the cardiac muscle.

- **Blood:** Transport medium containing plasma, white blood cells, platelets, red blood cells and many more.
-

Where is Blood Made?

Blood is made in the bone marrows of our body. The main source where red blood cells are made are from our stem cells, therefore, the millions of red blood cells made are taken by the blood vessels, also, the life span of these red blood cells from 100-120 days depending in the individual themselves.

Blood Components
The components of blood are listed in the table below:

Blood Component Table

Component Name	Role of Component
Plasma	Carries all cells around the body, followed by dissolved nutrients, hormones, carbon dioxide and heat.
Red Blood Cells	Carries oxygen around the body. Its role is also to unload oxygen to the muscles of the body.
Lymphocytes	Produce antibodies to destroy all harmful types of microorganisms. These cells help give the body immunity.
Phagocytes	Increase our immunity by the process phagocytoses.
Platelets	Releases chemicals for blood clotting, therefore, this is what heals our cuts.

Types of Circulatory Systems
There are a total of two circulatory systems, therefore, these two are:

- **Single Circulatory System:** This is where the blood is pumped goes around the body once; since it leaves the heart, reaches the lungs before finally going to the whole body. Animals that have this system are fish.

- **Double Circulatory System:** This is where the blood is pumped around the body twice, since it first leaves the heart, goes to the lungs, back to the heart, before finally going to the rest of the body.

Circulatory System Procedure
The circulatory system all begins with the heart. Therefore, to know how the circulatory system works; we need to understand how the heart beats and sends blood to all the organs in our body, as well as how blood goes to the lungs; before coming back to the heart. In our circulatory system, we have two circulatory circuits; which work at the same time called the pulmonary and systemic circuit. Therefore, it's better to see how our heart pumps blood, by looking at the procedure of the pulmonary and systemic circuit separately. Firstly, let's look at how the systemic circuit pumps blood from the heart, to the many organs of our body.

Heart Beat Names
- **Systole:** When the chambers of the heart is contracting, therefore, closing the bicuspid and tricuspid valves of the heart, this stops blood flowing into their atriums and ventricles.

- **Diastole:** When the chambers of our heart relaxes, therefore, opening of bicuspid and tricuspid valves of the heart to let blood though its atriums and ventricles.

How is our Heart Beat Regulated?
When one places their hand close to the middle of their chest, a heartbeat will be heard. This due to the systole and diastole functions working together and pumping after each other. The heart normally beats 70 times per minute; however, this can vary depending on the condition of the individual and what they are doing, therefore, the answer to this question; is that our heart rate is regulated by the medulla in our brain. When the body needs to increase heart rate, such as in aerobic exercise to rid carbon dioxide, sensors in the aorta, and the carotid artery, detect this; and send nerves impulses up to the medulla. Once the medulla receives these nerve impulses, it will

send nerve impulses down to the accelerator nerves; so the heart pumps more blood at a faster rate around the body. This also goes oppositely, since when the c02 level returns to normal, the medulla will receive fewer impulses, therefore, the medulla will then send nerve impulses to the decelerator nerves. This will decrease the heart rate and bringing it back to 70 beats per minute

Systemic and Pulmonary Circuit Procedure
Systemic Circuit Procedure
When our heart is in systole, oxygenated blood, made from our bone marrow, flows though the pulmonary veins; and enters the left atrium; however, the blood can't pass in the left ventricle yet; since the bicuspid valves will be closed. Once the heart is in diastole, the walls of the aorta will contract, raise the pressure in the atria, and makes blood from the left atrium flows into the left ventricle, therefore, this means that the heart will go into systole again and the valves will close to stop any blood from flowing backwoods. Once the heart goes into systole, the bicuspid valves will open, the pressure will rise; and the blood will flow into the aorta. When the blood is pumped out to the tissues, muscles and other organs of the body; the pressure in the heart will rise, the heart will go into diastole; and the pulmonary artery closes the bicuspid valves in the blood vessels.

Pulmonary Circuit Procedure
In the pulmonary circuit, when our heart is in systole, deoxygenated blood will flow into the heart from the vena cava, into the right atrium. When the heart goes systole, the de-oxygenated blood flows into the right ventricle; therefore, once the heart goes into systole again, the deoxygenated blood will flow from the right ventricle into the pulmonary artery, before it's pumped to the lungs to receive oxygen.

How do these two System Circuits Connect?
How these two system circuits connect, are to do with what happens to our red blood cells when they travel all around the body. Therefore, in the systemic circuit, when the blood pumps out of the aorta, though its arteries, to the body; the red blood cells release oxygen and nutrients all round the body, while taking in carbon dioxide and wastes from the bodies organs, tissues, and muscles. This is what turns of the oxygenated blood into deoxygenated blood.

Vein and Artery Differences
There are some well defined differences between the structure of the veins and arteries. Such differences are
- **Wall:** The walls of the artery are a lot thicker than the veins; which are a lot thinner.
- **Lumen:** The lumen of the artery is a lot smaller than the lumen of the vein; which is larger.
- **Lining:** The Lining of the artery is thinner than that of the vein; which is thicker.

Why are the arteries Larger than the Veins?
The reason to why the arteries of the heart are larger and thicker than the veins; is due to the size of either side of the heart. We notice that in our hearts, the left side of the cardiac muscle is larger than the right side; therefore, this is due to the left side pumping more blood to the muscles, and organs, of the body; than the right side; which only needs to pump blood to the lungs and back to the heart. Arteries are responsible for taking blood to all the tissues and organs of the body, therefore, they are larger and gain more muscle; due to the fact that they need to cope with the constant pressure of pumping blood; as well as allowing itself to stretch, when more blood is being pumped, and recoil; when there's not so much blood needing to travel around the body quickly.

Capillaries
How our body gives oxygen and nutrients, while taking wastes and carbon dioxide, is due to our capillaries which are all over our bodies. These capillaries; which are located in the brain, heart, arms, lungs, liver, organs of the digestive system, kidneys, legs, and reproductive organs, are the

sources where our red blood cells can give nutrients, while taking waste at the same time. Once our blood becomes deoxygenated, it flows from the organ it's given nutrients to, though its veins and back to the superior and inferior vena cava; where it is pumped though to the pulmonary artery, and sent to the lungs. At the lungs, the red blood cells will receive oxygen and become oxygenated blood, therefore, once the blood is oxygenated, the blood will flow back to the heart again, which means the pulmonary circuit is finished, and the systemic circuit will begin again; where the blood will be pumped though to the aorta, before being sent around the whole body.

Coronary Arteries
These arteries are what carry the red blood cells to the narrow capillaries to give oxygen and nutrients.

Immune System
What is the Immune System?:
The immune system is normally described as a network organised into subsets which regulate, pass information back and forth and protect the whole body from microorganisms. These pathogens carry bacteria that invade and harm the functions of the body. For example, these guards of the body are known as the skin, cells lysozyme, lymphocytes (white blood cells), stomach acid antibiotics and vaccinations. The immune system is one the complicated but amazing systems of the body due to its large memory bank which remembers all the infections it's ever encountered and has the ability to create secretion of certain cells in order to kill the infections.

What Type of Microorganisms can harm the Body?
There are four microorganisms that are capable of turning into pathogens and harm the body, if barriers of the immune system not strong enough, or are affected by the microorganisms. There are good types as well as bad types of microorganisms and these five are Fungi, Protoctisits, Bacteria and Viruses.
- **Fungi**: This microorganism has cell walls which are made from cellulose; they do not contain chloroplasts in their cells. In nature, fungi are also mushrooms, toadstools, puffballs and bracket fungi which can live anywhere such as in water, fruits and soil. Another type of fungi is called yeast; the yeast is unicellular and can thrive and grow anywhere that is warm and damp. For example, if a warm damp medium; such as bread or fruit is left out in the air, surface, or atmosphere for a few days, the yeast (from the air) will land on the food and start to absorb all the nutrients from the food. This makes mould grow in it before finally making the food inedible and dead, therefore, the process of this digestion is called sapotrophic nutrition. This mould is the fungi's fruiting, or growing body which underneath have thread like filament spores called hyphae which invade the surface of its food, grow hyphae braches over its surface, secrete extracellular digestive enzymes, which break the food down into soluble substances such as sugar and finally absorb the nutrients from the material or food it lands on.

How can Fungi Harm the Body?
In the body, yeast is naturally good for us, but, there are yeast related illnesses such as Athletes foot, Yeast skin rashes and Candida (Thrush). There are certain causes of these illnesses, for example, a person can get athletes food by the over growth of fungi called dermatophytes which then feed on the surface of the foot to stay alive. Canadidal yeast grows and feed on the surfaces of skin which can cause burning, itching and skin rashes. As for oral or vaginal thrush, the cause of this illnesses is when the over growth of candida albicans outnumber and take over the friendly bacteria which are located on the vaginal wall, throat, tongue or the inside of the cheeks. Unfortunately, this can cause the person to have itching, discomfort burning, inflammation and urine infection.

- **Protoctisits:** Protoctisits are a mixed group of eukaryote living organisms that can't be described as plants, animal or even fungi but are described as multi-cellular with a rigid cellular wall and make their own food though photosynthesis. These eukaryote microorganisms are mostly described as plant-like organism without the roots, stems or leaves, however, they are mainly aquatic and live in the water and there are many different types of these protoctists. For example, photoautotroph's are seaweed protoctisits which crate oxygen under water as well as others such as Chromistia, Plantae, Oomycota, which are water moulds, Apicomplexa, brown algae called Pheophyta, and green algae called Chlorophyta.

How can Protoctisits Harm the Body?
Protoctisits are the same as fungi; therefore, they feed and live of animals, plants and humans in their warm damp environments. Some protists are the causes and agents of disease for example, Plasmodium, which are protoctists. This is an organism that causes Malaria as well as Dysentery.

- **Bacteria:** Bacteria are single cellular living organisms with a cell wall, not made of cellulose but made from proteins sugars and lipids, they can do all the characteristics of a living cell. For example, they can grow, move, excrete, reproduce and eat; they have circular shaped rings of DNA that carry the bacteria's genes called plasmids. These plasmids are useful in genetic engineering, as well as having a slime wall that protects the bacteria's shape and DNA strand, cytoplasm and a chromosome consists of a single strand of DNA. Also, the shape of bacteria is normally rod-shaped or spiral shaped with filaments called flagella projecting from it.

How can Bacteria Harm the Body?
Before we go into how the harmful bacterium affects the body, it's important to know that there are good types of bacteria the body has and needs. For example, Lactobacillus Bulgaricus are live rod shaped bacteria that are used in the production of live yogurt from milk, they are used in the manufacture of proteins and there are types of bacteria that are very important decomposers because they break down dead plants, materials and animals so that their atoms and essential elements can be used by other living organisms. However, harmful bacteria are normally ones that feed saprotrophically on cells and are described as parasites rather than harmful bacteria.
The way these parasites harm the body, is that that surround their target cell, secrete substances that damage and inflame the cell before they enter the cell and starting multiplying and reproducing more bacteria. Once the process is complete, or if the immune system does not kill them quickly enough, the parasites will flow into the blood stream and affect other cells of body, until the person becomes ill and the immune system is triggered. The illnesses that come from an invasion of bacteria can be Cholera, Food poisoning, Tuberculosis, Meningitis as well as the excretion of harmful toxins, which are mainly excreted by Clostridium bacteria.

- **Viruses:** Viruses are the smallest types of microorganisms, they can have many types of structures but in their cell structure, they don't have any nucleus, cytoplasm, cell organelles or cell membrane. Instead they have a genetic material of DNA or RNA surrounded by a protein coat, which can be stolen by the host membrane and used as an envelope to surround the virus particle. Therefore, these microorganisms are not regarded as a species or a living organism by scientists, since they don't do the functions a cell would naturally do. For example, they can excrete, move, grow, eat or respire; therefore, the best way to describe a virus is that it is between a non-living chemical and a living organism.

How can Viruses Harm the Body?
Although viruses are not considered as living organisms, there is one function a virus can do which is a characteristic of a cell, but unfortunately, this is the process which causes harm and

affects the body; this characteristic is reproduction. Viruses want to live and reproduce more of its many kinds but they can't do this by themselves, they need a living host, or medium, to create more viruses. Therefore, they do this parasitically by first finding a living cell and sticking to its membrane. Once it has done this, the virus particles will then enter the cell and break down its protein coat so it's RNA, or DNA, is released.

The RNA, or DNA, then replicates and orders the living hosts genetic machinery to make more of the viruses particles, once many particles are created, the particles burst out of the host cell, thereby killing it before spreading to other parts of the cells and doing the same process again. Unfortunately, if the immune system is weak or does not destroy the virus quick enough, the viruses will cause permanent damage or serious illness, such as HIV aids(Meaning: Human Immunodeficiency Virus and Acquired Immune Deficiency Syndrome), which can cause the individual to die. Other illnesses that viruses can cause can be colds, HPV, Poliomyelitis, Measles, Mumps, Chicken pox, Herpes, Rubella and Influenza as well as causing death and illnesses in plants, for example, tobacco mosaic viruses that affect tomato plants.

How can Individuals catch these Pathogens?

There are a few ways people can create, catch and spread these pathogens before becoming ill. These possible ways are coughing, sneezing, breathing, physical contact, lack of hygiene and coming into contact with vectors which can give diseases.

How does the Immune System Work?

If we look at the immune system as a main networking building, we can see that there are four levels of defences in the immune system which contain different organs, chemicals and cells that protect, fight and defend the whole body from infections.

First Floor of Defence
Physical Barrier:

- **Skin:** The skin is the first layer of defence and is the largest organ of the whole body. Its functions are to regulate the bodies' temperature, prevent the loss of too much water, control the loss of heat through the skins surface and of course are one of main barriers, which block the entry of disease-causing microorganisms. The top layer of skin, called the keratin or epidermis, creates an effective barrier that stop foreign bacteria or viruses getting in, however, this top layer of skin can easily be damaged or removed, for example, scraping, bruising and cutting can be the causes. There are certain ways the skin does try and block the pathogens before they come into the blood stream and spread their disease, for example, clotting is when the damaged, or open skin, is clumped up by proteins called platelets. These platelets gather together and block the smaller capillaries from the pathogens, the platelets, followed by the damaged cells, also produce a substance called fibrinogen that changes into a network of fibres across the wound called fibrin. Finally the red blood cells underneath become trapped in the network before and form a blood clot which stops anymore bacteria entering the body, therefore, this how a scab is created.

- **Sweating:** The reason why some people have high temperatures, and go though vasodilatation when they have flu is because the immune system is trying to get rid of the foreign bacteria in the body. Therefore, the immune system will raise the temperature so that the person sweats and get rid of impurities, as well as the foreign bacteria itself.

Chemical Defences:

- **Eyes, Nose and Saliva:** These body parts secrete substances of tears, mucus and saliva. All three of these substances contain enzymes called Lysozyme that will break the wall of the foreign bacteria, therefore, killing it. Tears not only break down bacteria's cell wall but also wash out any dirt or germs that have landed on the eye. In the nose, inner lining of the gut and lungs, have epithelium linings containing tiny hairs called chemoreceptor's, at the end of these receptors, there is a find film called mucus. Therefore, this mucus is what stops airborne pathogens from entering the nose, going down the windpipe, and affecting the lungs.

- **Stomach Acid:** As we eat food, we ingest many different types of bacteria which can causes illnesses and affect the body, fortunately, most of these bacteria are killed by the stomachs acid. Our stomach lining produces hydrochloric acid which makes a weak gastric juice; therefore, the role of this acid is to kill many of the bacteria that are on our food. However, there's another question we may ask on this subject, and that is why hasn't our stomach been digested by the hydrochloric acid since this acid is capable of burning and digesting some metals!

The reason our stomachs are not digested by the acid, is because of the body's marvellous way of keeping everything safe. This means our stomach wall has a lining containing epithelial cells that secrete protective substances of mucus and bicarbonate on top of the first layer of the stomach wall, which is protein. The mucus provides a physically barrier, so the acid can't touch the protein wall itself, while the bicarbonate, which is highly alkaline and is secreted by the stomach cells, neutralizes the acid.

Second Floor of Defence

- **Inflammation:** Inflammation is the response the bodies' immune system when the body is under attack by pathogens. Inflammation can occur in any illnesses, for example, flu, chills, scar tissue, fatigue and headaches. Therefore, once the skin tissue is damaged, or the immune system starts the inflammation response; the process begins.

- **What Happens in the Inflammatory Process?**
  When the body is attacked by pathogens, due to tissue damage or catching a foreign pathogen, chemicals from the damaged tissue, such as histamine, prostaglandins and kinins, start to be released. This is to activate the inflammatory response, once it has done this, the chemicals work together to cause vasodilatation, where the capillaries are opened wide. Inside of these capillaries, there are chemo taxis which are attracted by the chemicals messages to the injured site. Some of the chemicals even increase the sensitivity of the damaged area by stimulating the nerve cells; therefore, the damaged area becomes painful to warn the person that they have been injured. Once the chemical Chemotaxis has been attracted to the injured site, the chemical starts to migrate and create white blood cells called leukocytes, or phagocytes, to the damaged area. There are two types of these leukocytes called macrophages and neautophills and the role of these two leukocytes is to neutralize bad bacteria and engulf harmful bacteria.

- **Auto-Immune Disease:** This disease can be passed down through the family and the cause of the disease is by the persons own immune system. The way the body can become damaged by the disease is that the immune system releases the phagocytes and instead of attacking the bacteria, they go and attack the body's own tissues. Unfortunately, damage of these tissues can cause the person to create their own diseases. Such auto-immune diseases are Multiple Sclerosis; where the myelin of the spine is attacked, Rheumatoid Arthritis; where the tissues of the hands or feet are attacked and Crohn's Disease; in which the wall of the stomach is attacked.

- **Phagocytoses:** Phagocytoses is the process where white blood cells engulf and ingest harmful bacteria, in order to keep the body protected. There are many different types of cells, but the cells that that are responsible for engulfing and ingesting harmful bacteria are white blood cells called phagocytes. If the harmful bacteria are able to break though the first floor of defence; and are still present, they would be known as an antigen, which then triggers the immune system to send out phagocytes. There are 70% of these phagocytes in the body that engulf and ingest harmful bacteria. Once they detect the harmful bacteria, they go and look for it by following the chemical products which are coming from the pathogen. Therefore, they won't stop looking, until they are at the site of the wound, or when the pathogen is located.

Once the phagocytes find the pathogens, whether in the body or in the blood capillaries, they start to change their shape and produce extensions of their own cytoplasm called pseudopodia. Once the pseudopodia are created, the phagocyte then goes and attaches to the pathogen using its variety of surface receptors such as lippolysaccharide, antibody and complement receptors. For example, a complement receptor, such as Cb3, coats the bacteria with Cb3 receptors before binding itself to the pathogens receptors. Therefore, this process is called optimisation; which increases the process of phagocytoses. Once optimisation is complete, the phagocyte then engulfs the pathogen into its vacuole called a phagosome. Therefore, once it's inside the phagocyte releases digestive enzymes to break the pathogen down and that's how the pathogen is killed. When the body recovers from the illnesses, the person has gained natural acquired immunity. This means the immune system is stronger and that it would be very unlikely for the person to catch and suffer illness again.

- **Lymphocytes:** Lymphocytes make up about 25% of the over phagocyte cells that are in the body and the role of these lymphocytes are to make chemicals called antibodies. These antibodies attack the antigens of the bacteria or they can attach to the antigen. To stop it in its tracks, they make themselves a label for the phagocyte or they cause the pathogen to burst open, so that the phagocyte can find and digest the pathogen easier. There are two different types of lymphocytes which are T lymphocytes and B lymphocytes.

- **T Lymphocytes:** T lymphocytes do not recognise free floating pathogens unless they are carried on the cell surface of the bodies on complex molecules. T lymphocytes have specialised antibody receptors which can notice other fragments of pathogens of infected or cancerous cells that have been created by viruses. The role of these T lymphocytes is to regulate immune response, while other T lymphocytes, will go and attack cancerous cells. There are special TH lymphocytes that regulate the immune system by communicating with other cells, stimulate B lymphocytes (to produce antibodies,) activate other T lymphocytes and call more phagocytes to come to the pathogenic area, to kill the pathogens. Finally, there are special Cytotoxic T lymphocytes that attack pathogens carrying harmful substances in their surfaces, and they can kill pathogens by growing inside of the infected cell.

- **B Lymphocytes:** The role of B lymphocytes is to change into memory cells and remember the pathogen which increases the bodies' immunity, creates the antibodies that are able to kill the pathogen, and makes sure that the same virus or bacteria cannot harm the body again. This is if the same pathogen came into the body and tried to affect it, the B lymphocyte memory of that pathogen will start to reproduce the same antibodies, at a higher level for it. Therefore, the pathogen will be killed much faster than the first time. Lymphocytes can also form into plasma cells, which secrete antibodies that attack the pathogens in the bloodstream. There are many types of B lymphocytes, for example, there's a type of B lymphocyte that will be created for killing cold pathogens while another will be made for killing flu. Also there are certain bacteria that cause certain diseases, therefore, when a certain B lymphocyte finds it pathogen, the B lymphocyte will become active and trigger other plasma cells to produce antibodies, for example, immunoglobulin, Igm, Iga, IgE and IgD.

- **Phagocye Relitves:** Apart from the phagocytes engulfing and ingesting the harmful bacteria, there are other types of phagocytes that kill the harmful bacteria too. For exmaple, reletives of the phagocytes are Macrophages; which rid the body of worn out cell and secrete monkines which are vital for the immune system. Granulocytes contains chemicals, such as histmaine, which destory pthogens and cause immflamation, Netrophills which break down pathogens, and Dendritic cells which stimulate T lymphocytes. And show antigens the T lymphocytes.

- **Healing Porcess:** There are four healing processes which take place after the inflammatory response. These processes are Collagenation Angiogenesis, Proliferation and Remodelling.

- **Collagenation:** In this process, the Macrophages work on clearing the damaged area and make space for the rebirth of new tissue, while fibroblasts, collagen making cells, start to produce a new collagen matrix which will create a new framework for the skin.

- **Angiogenesis:** In this second process, which can also be called revascularization, the damaged area starts to create new capillaries to return blood back to it. When the blood has reached the area, this is able to make the new tissues cells grow.

- **Proliferation:** In the third process, the time it takes is up to about four weeks. Therefore, during this process, the affected area is composing itself of specific cell tissues, muscle tissue and granulation tissue. The ways scars are formed are from the granulation tissue, if the granulation tissue does not remove itself in the four weeks of the proliferation process.

- **Remodelling:** At the final process, the cells of the new tissue start to combine into their surroundings by arranging their protein fibres in a way so that the new skin will be suited to the new forces of stresses that might be imposed upon it. This process is not short and it can take many years before the new skin is created.

Renal System
What is the Renal System?
The renal system, is also known as the uniary system, therefore, the renal system is a part of the two kinds of excretory systems, which rids bodily wastes and fluids

What is the Role of the Renal System?
The role of the renal system is to send our blood though a filtration process, in order to cleanse the blood and get out the impurities which are created inside of our bodies. The impurities in the blood stream end up as typical urine, which is stored in our bladders; before being excreted though the vagina or penis. Without the renal system, we would not be living, since we would not be able to rid the toxic fluid wastes; which are created in our blood stream.

Organs that create this System
- **Renal Artery:** This is where blood from the aorta of the heart, which has gone all around the body, comes to be purified.

- **Renal Vein:** The newly purified blood will flow into the right kidney and the renal vein, where it shall be pumped back into the vena cava of the heart, before being pumped to the lungs in order to support the respiratory system as well as the rest of the body systems.

- **Kidneys:** Each human and animal has a set of two kidneys. Therefore, these kidneys are the main base from taking in blood from the renal artery, to purify it.

- **Tubules:** Take on the final filtration processes of the urine. They made sure that important nutrients and water is reabsorbed back into the body, while the waste is carried to the ureter.
- **Ureter:** This is where the urine waste is carried and stored into the bladder.

- **Bladder:** This is where the urine is stored before excretion.

- **Sphincters:** regulates the contraction and relaxation of the urethra, therefore, stopping any urine from seeping out.

- **Urethra:** This is where the remaining urine is excreted.

System Procedure
Since the typical adult produces 1.5dm3 of urine every 24 hours, and every litre of it contains 40g of waste products, the renal system is always working to cleanse our blood and produce the best purified blood; in order to keep our bodies functions in order. Here is the complete process of the renal system.

- **1:** Blood from the heart is pumped out though the Aorta, before going all around the body. In this process, the red blood cells will be collecting all types of wastes, for example, carbon dioxide from the lungs, ammonia, nitrogenous waste, sodium chloride, phosphate, water, ions, urea and potassium. When the blood finally reaches the kidneys; it's passed though the left side of the kidney; in order to go though the purification process.

Blood Purification Process
If we look inside of the kidney, we will see the many different muscles and structures which work to purify our blood. Therefore, inside our kidneys there is the cortex, medulla, nepheron, pyramids, pelvis and ureter, which purify our blood of toxic fluid wastes; before sending the purified blood back into the renal vein. As for the urine wastes, they are sent though the ureter to be further filtrated; before the remaining waste is excreted though the urethra.

- **2:** In the blood purification process, the blood travels though segmental arteries, though the helix, to the cortex, before finally reaching the medulla; which contains small unit filtering arteries of the renal artery; called kidney tubules or nepheron.

How is the blood filtrated from the Kidney Tubules (Nepherons) to the Bowman Capsule?
To find this question out, we need to know how the blood is filleted from the kidney tubules to the bowman's capsule; therefore, since the filtration process starts from the kidney tubules, it would be best to explain it; starting from there. In the medulla, where the kidney tubules are; the all the blood, is filtrated by nepheron. There are over one million nepheron in one kidney, and each one has a tiny hollow cup of cells called the Bowman's capsule.

How is the blood filtered from the Bowman's capsule to the Bladder?

- **4:** The Bowman's capsule surrounds a ball like shape of capillaries called the glormeruli. Therefore, this means that the blood will flow from the kidney tubules, into the nepheron, along smaller arteries called afferent arterioles, and arrive at the glormeruli; where it the blood is filtered at high pressure speeds. A very small blood vessel in the nepheron will carry blood away from the glormeruli; though efferent arterioles; before and leading it to capillary networks which are the pertubular capillaries. These capillary networks surround other parts of the nepheron, therefore, since the blood pressure flowing from the small arterioles to the glormeruli is very high; it causes the bloods filtrate substances to be forced out of the blood, though the capillaries, out of the bowman's capsule and finally into bowman's space which is the middle space between the glormeruli and bowman's capsule.

- **5:** In the capillaries, the blood goes though filtration by being separated by three different types of cell layers. The first one is called the capillary wall, while the other is called the capillary wall, also, between these two there is a third layer called the basement membrane. This is the main layer which filters filtrate substances made from water, ions, and small molecules out of the red blood cells. However, in this process, called ultrafiltration; the basement membrane will not allow any large molecules or red blood cells to be filtered out with the filtrate substances.

- **6:** After the ultrafiltration process, the filtrate substances, that have entered bowman's space, is now called the glormeruli filtrate, therefore, this is the starting product of the production of urine. This urine will go from Bowman's capsule space to the entrance tubal, or what is called the proximal convoluted tubule.

Final Tubule Filtrations

Before the urine flows to the ureter, balder, and gets excreted though the urethra; it goes though the final filtrations process, in order to reabsorb any important nutrients, as well as regulate the concentration and condition of the urine. Therefore, the process where the glormeruli filtrate goes though the four tubules; for these important purposes; is called selective reabsorption. In the final filtration processes, the urine goes though four different types of tubules.

Proximal Convoluted Tubule

- **7:** In this the first tubule, called the proximal convoluted tubule, over 80% of the water, urea, and nutrients needed by the body; will be reabsorbed back into the blood; in the peritubular capillaries, and flowed back around the body by the renal artery. This means that 20% of the glormeruli filtrate will be left to flow to the second tubule called the Loop of Henle.

Loop of Henle

- **8:** The role of the Loop of Henle; is to change the concentration of the urine, by taking 99% of the water and ions away from the glormeruli filtrate (urine). To go into more detail, the way the Loop of Henle is able to do this, is by firstly carrying down the filtrate into the medulla. However, since the lower part of the Henle loop is semi-permeable, this causes the filtrate to pass into the medulla. The top part of the Henle loop then carries the filtrate upwards from the medulla, which in the process absorbs 99% of the water and ions into the body and therefore increases the concentration of the urine.

Distilled Convoluted Tubule

- **9:** The concentrated urine is then passed into third tubule called the Distilled Convoluted tubule; which regulates potassium, sodium and ph levels of the urine, followed by further filtration of the urine.

Collecting Duct

- **10:** Finally, once the levels of the potassium, sodium and ph are regulated, the filtrate finally passes to the fourth tubule called the collecting duct. This is where sodium levels are regulated followed by any last nutrients or needed water is reabsorbed back into the blood. The collecting duct is also known as the ureter, therefore, the remaining urine flows though here, enters the pelvis, bladder, and is finally excreted though the urethra.

- **11:** The purified blood will finally flow out of the peritubular capillaries to the inferior vena cava, where it is known as Venus blood, and will flow back to the heart. In the process of selective reabsorption, a process called tubular secretion can also take place, where any impurities or waste in the blood; can be secreted from the peritubular capillaries, into one of the four filtration tubules

Other Roles of the Kidneys

Apart from excreting urine to keep the body healthy, the other roles of the kidneys are to:

- **Regulate Body Water Content:** Our kidneys can also regulate the water content in our blood. This means that if we drink a lot of water, the kidneys will feel this upset in water levels, and will create more diluted urine to rid the excess water. If we have too little water, our kidneys will produce smaller levels of urine. This regulated is done by the kidneys releasing the hormones called ADH.

Endocrine System

What is the Endocrine System?

The Endocrine system is a coordination system which creates different types of hormones, before releasing them into the bloodstream of the human body. Once these hormones; also known as pheromones, are sent around the body, the specialised hormones reach their target organs and begin their functions. For example, hormones secreted by the endocrine glands control the tempo of the organs functions by speeding the activity up or slowing them down. They can also control the metabolic rate of all body cells, they can affect the growth rate of human body and bones, and they play a part in the secondary sexual characteristics of human beings, as well as regulate the amount of water reabsorbed back into the body.

Endocrine Glands

There are nine endocrine types of endocrine glands in our bodies. These glands are:

- **Pineal, Hypothalamus, and Pituitary Glands**
- **Thyroid Gland, Parathyroid and Thymus Glands**
- **Adrenal Glands**
- **Pancreas**
- **Ovary/Testis**

Endocrine Hormone Groups

Along with the five endocrine groups, all the hormones made from these glands have been separated to create three different types of hormone groups. These three hormone groups are:

- **Steroids:** Steroids are hormones which are made from fats called cholesterol. These types of hormones are normally secreted by the sexual endocrine glands, for example, the ovaries, adrenal cortex, placenta or testis. Examples of steroid hormones are Testosterone, Progesterone, Aldosterone, Estradiol and Androstenedione.

- **Peptides:** Peptides are made from amino acids and they are secreted from the pituitary gland, parathyroid, stomach, liver, and kidneys. Examples of these hormones are the growth hormone, Vasopressin, Thyroid Stimulating Hormone, and Oxytocin.

- **Amines:** Amines are made from the amino acid called Tyrosine and are normally excreted by the thyroid and the adrenal medulla.

Endocrine Glands
Pineal, Hypothalamus, and Pituitary Glands
What is the Pineal Gland?
The pine coned shaped Pineal gland is located in the diencephalon of the brain; which is near the middle of the whole brain. It whole structure is made from Pinealocytes cells and glial cells which are the cells of the nervous system. The Pineal gland is known to be the third eye in mysticism.

What is the Role of the Pineal Gland?
The Pineal Gland secretes a hormone called Melatonin ($C_{13}H_{16}N_2O_2$). Therefore, this hormones role in the body is to:

- **Regulates Sleep and Wake Cycles:** This is the reason why humans will be more active in the summer, and more tired in the winter. Melatonin also regulates animal's hibernation cycles in the winter. The reason why humans and animals will feel more tired in winter is because the hormone Melatonin is secreted mostly during the night because when we close our eyes, the Pineal gland senses that it's dark and starts to secrete the hormone. However, the secretion of Melatonin starts to slow down as the environment gets brighter. Therefore, since in winter the environment is darker, more of the hormone is secreted to the body; which means that the individual or animal will feel more tired. In humans, excessive amounts of this hormone are the cause of jet lag, weight gain, and sadness.

- **Influences Sexual Development and Reproduction**

- **Convert Nerve signals to Endocrine Signals.**

- **Repairs Damaged Cells:** Once the Pineal Gland secretes the hormone, it sends hormonal signals to all the cells in the human body to repair themselves.

- **Regulates Endocrine Functions.**

- **Influences Depression, Irritability and Suicidal Tendencies:** This is caused by extremely low amounts of Melatonin.

What is the Hypothalamus/Pituitary Gland?
The Hypothalamus is a pearl shaped gland which is located at the under surface of the brain below the Thalamus and controls the Pituitary Gland. The reason the Hypothalamus also controls the Pituitary Gland, is because the Pituitary gland is attached to the Hypothalamus, therefore, these two glands work together to produce and secrete many hormones that are needed by the human body. The Pituitary gland is located in the bony hollow of the skull called the Sella Turcica; which is located at the back of the nose and underneath the brain.

What is the Role of the Hypothalamus/Pituitary Gland?
Both these glands secrete a number of hormones which control the homeostasis in the body, also, both these glands also respond to the environment around them. The table below shows the different types of hormones secreted by the Hypothalamus and Pituitary Gland, followed by the function of the hormone.

Roles of Hypothalamus and Pituitary Gland Table

Hormone	Gland its Secreted By	Function of Hormone
Prolactin	Pituitary Gland and Hypothalamus	Induces milk production in the female breasts after child birth.
Adrenocorticotropic Hormone (ACTH)	Pituitary Gland	Simulates Adrenal Glands to release Cortisol; which is a steroid hormone that causes the individual to have stress.
Growth Hormone	Pituitary Gland and Hypothalamus	Regulates growth, body composition and metabolic rate.
Anti-Diuretic Hormone (ADH)	Posterior Pituitary Gland and Hypothalamus	Regulates the amount of urine which is created by the kidneys. It also regulates the amount of water which is absorbed in the kidney tubules.
Oxytocin	Pituitary Gland and Hypothalamus	Controls the contraction of the womb in pregnancy; as well as induces milk for the baby after child birth.
Luteinising Hormone, Follicle Stimulating Hormone (FSH), and (Gonadotrophins)	Pituitary Gland	Stimulate sex hormone production as well as mature eggs, produce oestrogen, induces ovulation, creates the corpus luteum, and matures sperm.
Melatonin	Pituitary Gland and Hypothalamus	Controls sleep and awake cycles and the change in Cortisol.
Thyroid Stimulating Hormone (TSH)	Pituitary Gland	Stimulates the Thyroid Gland to secrete Thyroid hormones.
Endorphins	Pituitary Gland and Hypothalamus	Creates the feeling of pain.

Thyroid Gland, Parathyroid, and Thymus Glands

What is the Thyroid Gland?

The Thyroid is a butterfly shaped endocrine gland; which is located in front of the windpipe at the front part of the neck. They thyroid is controlled by the Hypothalamus and the Pituitary Glands.

What is the role of the Thyroid Gland?

The role of the Thyroid is to produce the hormones called Thyroxine and Triiodothyroine, The role of these two hormones is to regulate the metabolism of the heart, muscles, all body cells, lungs, and other organs of the body. Other roles of these two hormones are to promote natural skeletal health, regulate body temperature, regulate weight, as well as being important in the development of the brain. In order for any of these functions to happen, the gland needs iodine to make the hormone. Therefore, since Triiodothyroine has four ions of iodine, the thyroid can create the two hormones needed for the functions. Before the functions take place Triiodothyroine is converted to Thyroxine; therefore, due to this hormones activeness being more active than Triiodothyroine, this is what influences the cells of the body to do their functions. Calciitonin, a hormone also secreted by the thyroid, stops oesteoclasts from happening in the bone marrow.

Parathyroid Glands

There are four mustered coloured pea sized parathyroid glands; which are located a little bit behind the actual thyroid gland with two glands lying behind the butterfly shaped wing. When the thyroid senses something wrong in the body; for example, high levels of calcium, it will secrete a hormone called Calciitonin. This will flow though the blood to the bones and stop the activity of oesteoclasts in the bone marrow. This action causes the decrease of calcium in the blood and stimulates the parathyroid glands to release their hormones.

What is the role of the Parathyroid Gland?
The role of the parathyroid glands to secrete its hormones, therefore, its hormones helps to:
- **Release Calcium from bones.**
- **Absorb Calcium from the intestine into the blood.**
- **Takes back Calcium located in the urine.**
- **Stimulates kidneys to excrete phosphate in the urine.**
- **Increase magnesium levels in the blood.**

Calcium
Calcium is very important, especially for the nervous system. The reason for this is because the role of Calcium is to provide electrical energy for the nervous system as well as for the muscles of the body. Regulation of calcium in the body is the only role that the parathyroid does.

Thymus
What is the Thymus?
The thymus is a gland located between the upper part of the breast bone and the heart. It has two lobes which are made from lymphoid cells. The thymus is normally at a large size until the individual hits puberty, therefore, when he or she does, the thymus will start to shrink.

What is the Role of the Thymus?
The role of the Thymus is to secrete hormones called Thymuses. These hormones stimulate the body to create T lymphocyte cells; which are needed by the immune system. Also, these hormones help the body create its own lymphoid system.

Adrenal Glands
What are the Adrenal Glands?
Adrenal glands are endocrine glands that are located on top of the kidneys. The outer part of the adrenal glands is the adrenal cortex; while the inner part is the adrenal medulla. The adrenal medulla is made from Chromaffin cells; while the adrenals cortex is made from a range of cells, for example, Glomerulosa, Fasiculata and Reticularis cells.

What are the roles of the Adrenal Glands?
The adrenal glands are mainly responsible for the fight or flight response. This is to the adrenal medulla secreting the hormone adrenaline. Therefore, this causes the individual to become scared, very happy, or angry. Adrenaline also causes the individual to move faster and act quicker in situations that can be a threat to their lives. The table below shows the different hormones excreted by the adrenal glands; followed by their roles.

Role of the Adrenal Glands Table.

Hormone	Gland its Secreted By	Function of Hormone

Cortisone, Cortisol, and Corticosterone	Adrenal Cortex	Combats Inflammation, produces sugar glucose from amino acids, utilizes fat, protein and carbohydrates. The hormones also regulate salt and water in the body.
Adrenaline, Epinephrine and Norepinephine	Adrenal Medulla	Adrenal triggers happiness, anger, or fear in the human body. Therefore, this individual will react to adrenaline and rather react or escape. This is the flight or fight response is. Effects of these hormones are increased heart rate, increase of deep breathing, muscle power improves and metabolic rate is increased.
Mineralcorticoids	Adrenal Cortex	Maintains and regulates electrolyte balance.
Glucocorticoids	Adrenal Cortex	Creates the effects of stress at a low pace. For example, such effects of stress, caused by the hormones, are high blood pressure, inflammation and disrupt the immune response.
Noradrenalin	Adrenal Medulla	Increases blood pressure, increases blood flow throughout the body, and increases breathing rate.
Estrogen and Testosterone	Adrenal Glands	Stimulates sexual development in boys and girls.
Dehydroepiandrosterone sulfate (DHEA-S).	Adrenal Glands	Their function is unknown; however, individuals with high amounts of this hormone heal quicker from illnesses, flu and colds.
Androgen Testosterone	Adrenal Glands	Takes part in fat, carbohydrate and protein metabolism.

Pancreas
What is the Pancreas?
The pancreas is an endocrine gland, exocrine gland, and digestive gland. Its location is behind the stomach. Although it's mainly known to secrete digestive enzymes; it also secretes glycogen and insulin to keep the body's sugar levels balanced.

What is the role of the Pancreas?
In the endocrine system the role of the Pancreas is to secrete Insulin and Glycogon. The table below shows what functions these two hormones have in the body:

Table

Hormone	Function
Insulin	Lowers and balances blood pressure; by breaking down the Glycogon to glucose.
Glycogon	Increases and balances blood pressure by decreasing the amount of glucose in the bloodstream.

Exocrine and Endocrine Glands

What are the Differences?
Exocrine glands are duct glands; which secrete their hormones though ducts to the outside of the body. However, Endocrine glands are ductless glands which secrete their hormones into the bloodstream. Examples of Exocrine glands are the tear ducts, mammy glands, salivary glands and anal glands, while examples of endocrine glands are the Pituitary Gland, Thyroid Gland, Ovaries/Testis Glands and Adrenal Glands. The only organ in the body, which can do both of these functions; is the Pancreas.

Ovary/Testis
What is the Ovary/Testis?
The Ovary and Testis are both sexual glands. These glands secrete sexual hormones to change the individual from a girl to a woman, or a boy to a man. The two testes are shaped like sacs underneath the penis and it has an outer layer of skin which covers a layer of muscle tissue. As for the ovaries; they are almond shaped glands, they lie on both sides of the uterus and are below the opening of the vagina where the fallopian tubes are.

What is the Role of the Ovary/Testis?
The role of these two sex glands is to secrete sex hormones. The hormones, followed by their location and functions, are in the table below.

Hormone	Gland its Secreted By	Function of Hormone
Testosterone	Testis's	Helps in the growth of the penis facial hair, bodily hair, breaki depending of the voice, incre muscle mass, increase of musc and height.
Inhibin	Ovaries	Stop the secretion of FSH and c egg development.
Oestrogen Progesterone, and Estrogen	Ovaries	Causes the woman to experie secondary sexual characteristi example, the breasts will deve menstrual cycle will begin a regulated, bodily hair will start t there will be an increase in hei voice of the girl will deepe fallopian tubes, uterus and vag mature, and the hormone will r the distribution of fat in the body
Prostaglandins	Ovaries	Aids in the variety of female and the sex hormones.

Nervous System
What is the Nervous System?

The nervous system is the most important system in our body. The reason for this; is because the nervous system carries thousands of different impulses around our bodies, so we can move, eat, sleep, and keep ourselves safe from danger.

Stimulus and Response

All responses start with a stimulus. Therefore, a stimulus, is something that causes impulses to travel though our neurons, flow into the spinal cord, flow though synapses; which go up to the brain; before the brain processes the information; and sends new impulses back down the spine, synapses, and neurons; to initiate a response in the body. Before the stimulus causes a response; the impulses go though three other processes; which are receptor, coordination and effecter. The nervous system events go in this order:

Stimulus, Receptor, Coordination, Effecter and Response

Receptors of the Body

Receptor	Type of energy processed to Chemical energy
Eye	Light Energy
Ear	Sound Energy and Movement Energy
Tongue	Chemical Energy
Nose	Chemical Energy
Muscles	Movement and Heat energy
Skin	Movement and Heat Energy

Organs that create the Nervous System

- Spinal Cord
- Brain
- Synapses
- Neurones
- Receptors
- Muscles

How Does the Nervous System Work?

Reflex Arc

The central nervous all begins; when the individual becomes in contact with a stimulus. Therefore, a good example of a stimulus; is when a person touches a sharp object such as glass, pin, or something that is extremely hot or cold. When a person touches something sharp, hot, or cold; and temperature receptors on the fingers will detect this stimulus and generate impulses in the sensory neurons.

1. Sensory Neurons

Sensory neurons are nervous cell which carries the generated impulses from the area the stimulus was detected; before flowing though junctions to the first part of the central nervous system called the spine. As the impulses flow to the spine; the impulses travel through long fibre extensions called axons. The axons are covered by a sheath made from fatty material called Myelin, therefore, the Myelin Sheaths role is to insulate the axons which prevent short circuits from happening. Also, their other role is to make the conduction of impulses faster, since their travel rate, between 10 and 100m/s; is much slower than an electric current. The sheath is formulated by special membrane cells; which will embrace themselves around the developing axon. Connected to the axons are cell body's; which have extensions called dendrons. These dendron's lead to smaller branching junctions called dendrites, therefore, the role of these dendrites is to carry impulses and make contact with other neurones in the body, for example the motor neurons.

Other parts of the Neurones

- **Node of Ranvier:** These parts of the neurons are the 1 micrometer gaps between the myelin sheath cells. The role of these nodes is unknown; however, they help prevent the decay of the nerve pulses, by amplifying them. The nodes have high amount of sodium chemicals, therefore, they are very important; since they are crucial to the transmission speed of impulses to and from the brain.

- **Schwann Cells:** These types of cells embrace themselves around the nerve axons of the neuron. Apart from being the structure of neurones; their other roles are to rid debris and waste from the nervous system, insulate the axons, and regenerate new axons.

- **Axon Terminals:** These terminals are where the impulses, from one neuron, pass to other neurones in the nervous system.

2. Spinal Cord

Once the impulses reach the spinal cord of the central nervous system; they first enter a spinal nerve called the dorsal root. The dorsal root describes the back part of the spine which is faced at the back of the person. Once the impulses enter the dorsal root; they go though special synapses to rather reach the brain, or reach the relay neurons; which lead to the motor neurons. In the middle of the spinal cord; there are cell bodies which are grey, therefore, this area is known as grey matter, while the outer part of the spine is known as white matter; due to the cell bodies being white in colour, as well as containing more axons with fatty myelin sheaths.

Structure of Spine

The structure of the spinal cord is divided into two roots called the Dorsal and Ventral roots. The Dorsal Root is where all the sensory impulses enter to reach the brain, while the Ventral Root is the root where impulses, from the brain, travel out to the motor neurons. Both sensory and motor neurons are contained in both root structures of the spinal cord, also, in the dorsal root, there are bulges; containing many sensory fibres galled ganglions.

Connecter Neurones

The connecter neuron is where impulses, two and from the brain, are transmitted to many other motor neurons, and synapses, in the body.

3. Synapses

Synapses are connecter junction; which are made from thousand of neurons. They are located in the spinal cord and brain. These synapses have gaps between themselves and other nerve cells. Therefore, when the impulses reach the synapse; it triggers the fine ends of the nerve cell axons to salivate a chemical called neurotransmitter. The role of this chemical; is to create a bridge between the synapse and the second neurone, therefore, impulses in the second cell will begin to send the same message from the synapse, to the spine, before reaching the brain. After the impulses have been passed on, the bridge is broken down by specialised enzymes.

4. Brain

Once the impulses reach the brain, they are sent to different parts of the brain. The brain is made from nerve cells bodies; therefore, this is why the brain is sometimes called grey matter. Axons are located in the middle of the brain; therefore, this is what makes the spine and brain different. The part of the brain; where all impulses from all our organs, as well as processed into information; is the Cerebrum; which is made from two cerebral hemispheres. In this part of the brain, all of our thoughts, pictures and memoires are kept; also, the outer layer of the cerebrum is called the cerebral cortex. This cortex covers the cerebrum to create it surface. The role of the cerebrum is to receive and process information from all organs; before sending them down from the motor areas of the brain; to the motor neurons. Other roles of the cerebrum; is to hold the needed functions of memory, emotion, and personality.

5. Motor Neurons

Once the impulses have been processed into information, and understood by the brain, the brain will send new impulses from the cerebrum to the motor neurons. These new impulses will go though the relay neurones in the spine, before flowing though the motor neurons in the ventral root of the spine; which finally reaches the muscles and glands. Once the impulses reach the glands and muscles; the muscles will respond to the impulses and give a reaction. In the case of the individual coming in contact with a sharp, hot, or cold object. The individual will move their hand quickly in order to avoid the danger and the possibility of being harmed and damaged. The process of the impulses travelling to and from the brain, to the muscles and organs of the body; is called transduction. The process of the impulses moving to brain, before new impulses are sent down to the motor neurons, happens in as little as three seconds, therefore, this information alone, followed by how the impulses travel, show how powerful and quick the central and peripheral system really is.

Interneurons

These neurons are nerve cells, which allow efficient neurons, afferent neurons, motor neurons and sensory neurons to communicate and send impulses with each others. Its typically know as the Multipotent neuron; with multiple dendrites, to send impulses to many different neurons of the central nervous system.

Peripheral System

This system is separated into two different systems called the somatic and automatic systems.

Voluntary (Somatic) and Involuntary (Automatic) Nervous Systems
What are the Differences between these Two Systems?

The differences between these two nervous systems are shown in the table below.

Somatic Nervous System	Automatic Nervous System
Voluntary actions are actions which we do that are under our control. With this system, we are able to do reflex actions consciously, for example, walking, talking, touching, eating, and thinking are all actions which are in our conscious control in the somatic nervous system. The somatic nervous system mainly controls the skeletal muscles and external sensory organs.	Involuntary actions are when we do reflex actions which are not in out conscious control. For example, blinking, peristalsis, beating of the heart, and the chattering of our teeth when we are cold. The automatic nervous system mainly controls the cardiac muscles and smooth muscles; such as the oesophagus in the throat.

Nervous System Event Table

Stimulus	Receptor	Receptor	Coordination	Effecter	Response
Bright light	Eye	Retina	Spinal Cord and Brain	Iris	Iris closes to shut out light.
Loud Music	Ear	Cochlea	Spinal Cord and Brain	Muscles	Cover ears.
Lemon	Tongue	Taste Receptors (Bitter, Sweet, Salty etc.)	Spinal Cord and Brain	Tongue	Likes or Dislikes taste.
Bad Smell	Nose	Olfactory Epithelium	Spinal Cord and Brain	Muscles	Cover nose from bad smell. The individual can also

					walk away from the bad smell.
The individuals hand is placed in Hot water	Skin	Temperature and Pin Receptors	Spinal Cord and Brain	Muscles	Takes hand out quickly to stop pain.
Sees something dangerous or scary.	Muscle	Stretch Receptors	Spinal Cord and Brain	Muscles	The individual would rather freeze in fear or run.

The Similarities and Differences between the Nervous and Endocrine System

Similarities	Differences
Both systems send messages to the whole body, in order to keep all functions active and healthy.	Reactions in the nervous system happen quicker and faster. Hormonal reactions happen slower in the endocrine system. Nervous system effects die out a lot quicker then hormonal reaction, which last much longer.
Both System produce Reactions.	The nervous system sends messages by impulses, however, the endocrine send messages by hormones
Both systems depend on the release of a chemical in order to send messages around the body.	The Impulses travel through neurons, the spine and brain. As for the endocrine system, it travels though the blood stream.
Both regulate the activity of cells, as well as the organs and tissues. With both systems regulating this, this means that both aim to maintain the homeostasis of the human being.	Hormones are regenerated, whole impulses are not regenerated.
Both systems use the same chemicals to send messages. For example, the hormones Epinephrine and Norepinephrine are neurotransmitters and are also secreted by the adrenal glands.	The endocrine system secretes hormones, while the nervous system sends impulses through the neurons.
Both systems are regulated by negative feedback mechanisms. Negative feedback mechanisms that return the body back to an unharmed state and produce counter-responses.	The endocrine system is a chemical reaction producer, while the nervous system is a physical reaction producer.

Reproductive Systems
What are Reproductive Organs?
Reductive organs are the organs which take part in the fertilization and reproduction of another human or animal species. All mammals have reproductive organs so that they can reproduce their species and carry on their generation.

Male and Female Reproductive System
Organs of the Male Reproductive System and their Functions
Organs of the male reproductive system are controlled by the hormone Testosterone. Therefore, these males' reproductive organs, followed by their functions, are listed below.

1. **Testes:** The testes are the home where sperm is made.

Sperm Production
In the testes, there are sperm producing tubules to which sperm is created. Sperm are cells which have goes through cell division, and have grown their own tails in the process, therefore, once the sperm cells are made; they pass into the epididymis. In the process of ejaculation, the epidermis and sperm ducts will contract to squeeze them out.

2. **Scrotum:** These are two special sacs; which lie on the outside of the abdominal cavity; also, inside of the sacs, they hold the testes, epididymis, and sperm duct. The scrotums are kept under the temperature of 37.3°c, therefore, this is the right temperature for sperm to be created and secreted.

3. **Epididymis:** This organ is a 6m coiled organ; which lies on the outside of the two testes. The testes are joining to the epididymis; as well as connecting to a muscular like tube called the sperm duct.

4. **Sperm Duct:** The sperm ducts are a tunnel that lets the sperm travel from the epididymis, through its duct, before reaching the small and coiled seminal vesicle branches.

5. **Seminal Vesicle Branches:** This is where semen is secreted, to collide and mix with the sperm, just before the man goes through ejaculation. The semen made from this contains fructose, therefore, this gives sperm the energy to travel through the uterus to the egg at the womb.

6. **Prostate Gland:** This surrounds the males urethra, as well as where sperm enters before going to the penis.

7. **Penis:** This is where the male's urethra is located; as well as where the male urinates, and ejaculates; during excretion or intercourse.

Erectile Tissue: This erectile tissue covers the penis, causes ejaculation to happen, as well as cause erections.

Organs of the Female Reproductive System and their Functions
Organs of the female reproductive system are controlled by the hormone Progesterone and Oestrogen. Therefore, these females' reproductive organs, followed by their functions, are listed below.

1. **Ovaries:** This is where the eggs, for fertilization, are produced before ovulation begins. The ovaries are 3-4cm long and one lies in the lower half of the abdomen at each side of the uterus.

Ovulation
The hormones called (FSH), and (LH); start stimulating the follicles; which makes them start to mature. In the maturing process, one of the female's eggs will grow inside the balls of follicle cells. Once the egg has matured; the follicle cells will start secreting a hormone called Estrogen and Oestrogen. Therefore, this hormone causes the uterus to repair itself, thicken itself; as well as the texture of cervical mucus to change and thicken. Another role of Oestrogen is to stop the secretion of Follicle Stimulating Hormone, while stimulating the secretion of the Luteinizing Hormone.
Once the Oestrogen has done its job; the Hypothalamus will release another hormone called Luteinizing Hormone releasing Factor (LH-RF), therefore, signalling the Pituitary gland to release another hormone called Luteinizing Hormone (LH). The presence of the LH hormone in the blood;

causes one of the mature follicles to burst; thereby, releasing its egg and setting it free. This process is called ovulation.

2. **Oviduct:** Narrow tubes that open the uterus and womb. Its length is normally 80mm long if there is no embryo present there.

3. **Fallopian Tubes:** This is where the released egg will travel down to the womb. Another name for this area is the oviducts.

4. **Cervix:** This area is rings of circular muscles, located at the lower end of the uterus where it joins with the vagina.

5. **Vagina:** A muscular tubule which is the outside part of the cervix and uterus.

Puberty
What is Puberty?
Puberty is a process in a human's life, where they start to change from a child to a teenager, before maturing into an adult, in this process, when the boy or girl is from the age of 10-14 years old; they shall experience changes in their body called the secondary sexual characteristics.

Effects of Puberty on the Sexes
Male
The secretion of Testosterone will cause:
- **Testis starts to produce sperm.**
- **Size of testis increases.**
- **Voice breaks and becomes deep:** This is caused by the boy's larynx, in puberty, to grow larger and larger; until it sticks to the apple Adam in the throat.
- **Pubic hair grows around the armpits, face, chest, and penis.**
- **Growth increases.**
- **Acne.**
- **Mood Swings.**

Female
Release of Progesterone and Oestrogen causes:
- **Increased breast tissue (This forms the girl's breasts).**
- **Hips widen.**
- **Pubic hair grows around the bikini area and armpits.**
- **Voice deepens and matures:** This is the same cause as the males voice breaking, however, the larynx doesn't grow as large as the males; which is why a woman's voice sounds more mature than deep.
- **Size of uterus and vagina increases.**
- **Menstrual Cycle.**
- **Acne.**
- **Mood Swings.**

How do Hormones control the Menstrual Cycle and Pregnancy?
To find out the answer to these questions; we need to find out what they are, followed by which hormones take part in them. Firstly, let's look into the menstrual Cycle.

What is the Menstrual Cycle?

The menstrual cycle is a period where the woman, or girl; goes though their reproductive cycle, and loses their placenta needed by the egg for fertilisation. If no fertilisation occurs by the time the egg ovulates, travels down the fallopian tube, and resides in the womb of the uterus; the placenta, egg, and corpus luteum; breaks down and flows though the uterus before coming out of the vagina. A period can last up to two weeks or a month; depending on the woman or girl. However, once the period ends; its takes 28days for the reproductive organs, followed by the sex hormones, to prepare themselves again for fertilization or for the next menstrual cycle. A period continues from teen hood, from the age of 12, to senior hood, where the women are passed 50 and go though the menopause process.

How does the Menstrual Cycle Happen?

1. Firstly, the Hypothalamus secretes the hormone called FSH-RF; which stands for Follicle Stimulating Hormone Realising Factor. Once this hormone has been secreted; it flows to the Pituitary gland to signal it to secrete the hormones called Follicle Stimulating Hormone (FSH) and Luteinizing Hormone (LH).

2. Once these hormones are secreted, and flow into the bloodstream, they start stimulating the follicles; which makes them start to mature. In the maturing process, one of the female's eggs will grow inside the balls of follicle cells. Once the egg has matured; the follicle cells will start secreting a hormone called Estrogen and Oestrogen. Therefore, this hormone causes the uterus to repair itself, thicken itself; as well as the texture of cervical mucus to change and thicken. Another role of Oestrogen is to stop the secretion of Follicle Stimulating Hormone, while stimulating the secretion of the Luteinizing Hormone.

3. Once the Oestrogen has done its job; the Hypothalamus will release another hormone called Luteinizing Hormone releasing Factor (LH-RF), therefore, signalling the Pituitary gland to release another hormone called Luteinizing Hormone (LH). The presence of the LH hormone in the blood; causes one of the mature follicles to burst; thereby, releasing its egg and setting it free. This process is called ovulation and sometimes women and girls can feel the burst of the follicle, however, this can be rare. The released egg will flow though the fallopian tube until it reaches the uterus, thereby alerting the body to prepare itself for pregnancy.

4. The remains of the burst follicle transform into the corpus luteum. The role of the corpus luteum is to secrete the hormone called Progesterone. Therefore, this completes the development of the uterus lining and thickens it even more for the fertilized egg, thereby, creating the placenta of the uterus. The hormone Progesterone maintains the lining of the uterus, develops more blood vessels for the placenta, and also signals the pituitary gland to stop the secretion of FSH and LH; which end the process of ovulation.

5. If the egg is fertilized by a single sperm, the egg will join to the placenta and start to form the baby. However, if the egg is not fertilized for 24 hours; the secretion of Oestrogen and Progesterone will stop. This is caused by the breakdown of the corpus luteum. Therefore, once the corpus luteum breaks down; the placenta of the uterus breaks down, hormones levels will drop, and the menstrual cycle begins. During this cycle, the placenta, corpus luteum and egg will shed and bleed though the uterus before coming out of the vagina. Once the menstrual cycle ends, it will repeat the preparation process again. This is due to the high amount of FSH and LH still present in the body; as well as the possibility that the next egg will be fertilized next time.

Summary of the Menstrual Cycle in Days

Day	Menstrual Cycle Action
Day 7	Hormones FSH and LH start stimulating the bags follicles.
Day7-Day 14	Hormones Estrogen and Oestrogen repair the uterus and thicken it.
Day 14	LH hormone causes the follicle to burst and send the egg down the fallopian tube. In the ovulation process, it also alerts the body to prepare for pregnancy.
Day 14-17	Egg flows from the burst follicle to the fallopian tube.
Day 17-25	The hormones Progesterone are released to do their functions. If the egg is not fertilized, the secretion of this hormone, followed by Oestrogen, will stop. The corpus luteum will also break.
Day 26-28	The placenta, corpus luteum and egg will break down before the menstrual cycle begins.
Day 28	Menstrual cycle begins where the placenta, corpus lutem, blood, and the egg will bleed out of the vagina though the uterus.

Summary of how Hormones control the Menstrual Cycle

Hormone	How it Controls the Menstrual Cycle
Follicle Stimulating Hormone Realising Factor	Signals Pituitary Gland to secrete Follicle Stimulating Hormone (FSH) and Luteinizing Hormone (LH).
Follicle Stimulating Hormone (FSH) and Luteinizing Hormone (LH).	Stimulates and Matures Follicles.
Estrogen and Oestrogen	Causes the uterus to repair and thicken itself. It also causes the cervical mucus to thicken. Another is that it stops the secretion of Follicle Stimulating Hormone, while stimulating the secretion of the Luteinizing Hormone.
Luteinizing Hormone releasing Factor (LH-RF)	Signal the Pituitary Gland to secrete Luteinizing Hormone (LH). Another action is causes, is of one of the mature follicles bursting; thereby releasing its egg.
Progesterone	Completes the uterus development by thickening the cervical mucus to create the placenta.

Why do Girls/Women bleed during their Period?

The bleeding in the menstrual cycle is due to the placenta peeling, or shedding, away from the uterus wall. Since blood vessels are connected to the placenta, due to the placenta needing the hormones to thicken, as well as carry nutrients and food to the fertilized egg; the shedding of their placenta causes the blood vessels to break and bleed. The amount of blood, which is normally lost during a period, is less that 80mls, however, heavy periods can cause the female individual to lose more blood than the average loss.

Effects of Periods

Such effects of periods on the individual are:

- **Increase of Irritability, Pain, Depression, Sadness, Grumpiness, Anger, Annoyance, Sensitivity and Depression.**

- **Anaemia:** Due to large amount of zinc; this is lost in the blood.

- **Cramps:** A period cramp, also known as Dysmenorrhoea, happens with the old uterine lining starts to break down inside of the body. In this process, hormones called prostaglandins are released into the body, therefore, causing the uterus muscles to contract. This contraction of the uterus muscles causes vasoconstriction to the blood supply flowing to the endometrium, therefore, once this endometrium dies, due to lack of oxygen, the uterine walls contact to squeeze this dead tissue though the cervix and out of the vagina. A inflammatory substance called Lukeotrienes are also released in the body and can cause menstrual cramps.

- **Back Pain**
- **Weakness**
- **Tiredness**
- **Bloating**
- **Craving for sweet foods = weight gain.**

How does Pregnancy Happen?

When an egg is fertilized, it will take a week to travel though the fallopian tube; before attaching itself to the uterine wall known as the womb. The egg will also go though its first mitotic division; before forming into the first stage of a baby called the blastocyst. This blastocyst causes more Progesterone hormone to be secreted and sent around the body; therefore, this creates a thick tissue between the egg and the uterine wall called the placenta. The placenta will transport oxygen and nutrients to the baby, so that it will continue to go though mitosis and form into a human baby, also, during this process, the placenta secretes a hormone called the Human Chorionic Gonadotropin hormone; which signals the body to support this pregnancy and not to abort the growing fetus. Another role of this hormone; is to prevent the decay of the corpus luteum. At the birth stage; nine months after the fetus has been in the womb, the placenta will start to secrete enormous amount of a hormones called Corticotrophin-releasing hormones; therefore, this stimulates the pituitary gland to secrete the Adrenocorticotropic hormone. This hormone, also known as ATCH, acts on the adrenal glands and causes them to secrete DHEAS; which is called Dehydropiandrosterone sulphate. The placenta will convert this sulphate to the hormones called Estrogen, therefore, this forms gaps junctions and create connexions. As labour begins; receptors for the hormones Oxytocin are secreted by the posterior pituitary gland, prostaglandins are synthesized in the uterus and placenta, before finally the uterine contraction is switched off and the uterus contracts to let labour begin.

After Birth

The hormones that take part after the baby is born are:

1. **Prolactin: Stimulated by the pituitary hormone.**
2. **Oxytocin: Stimulates milk release.**
3. **Inhibitory Peptide: Inhibits milk production.**

Summary of How Hormones Control Pregnancy

Hormone	How its Controls Pregnancy
Progesterone	Creates the thick tissue called the placenta.
Human Chorionic Gonadotropin hormone (hCG)	Alerts the body to support the pregnancy and not abort the baby. It also prevents the

	decay of the corpus luteum.
Corticotrophin-releasing hormones (CRH)	Stimulates pituitary gland to secrete ATCH.
Adrenocorticotropic hormone (ATCH)	Works on adrenal glands to secrete DHEAS.
Dehydropiandrosterone sulphate	Created to be converted onto the hormone Estrogen.
Estrogen	A form gap junctions and creates connexions.
Prolactin	Stimulates milk production.
Oxytocin	Causes prostaglandins to synthesis and uterine contraction. This hormones is what triggers labour to begin. It also causes the breasts to release the milk.
Inhibitory Peptide	Stops Milk Production

Fertilization: The process of fertilization only takes a day. Therefore, after 24 hours doctors will inspect the petri dish to see if any eggs have been fertilized. If the eggs have been fertilized, they will be left for five more days; before the fertilized eggs and taken are transferred back into the womb.

How does Fertilization Work?
After the reflex action of ejaculation, in which the male ejaculates his sperm into and on top of the vagina; the sperm, or spermatozoa, travels, or swims; through the cervix and up into the uterus using the wiggling movement of its tail. Once they enter the uterus, they then travel through the over duct and travels to the mature egg in the womb. Once the sperm finds the mature egg, one sperm will burrow at the shell of the egg, until its breaks through into the cytoplasm of the egg. The male nucleus of the sperm fuses with the female nucleus of the egg; this is where fertilization takes place. In an ejaculation, as many as five million sperm are sent to the egg; however, on their way to the egg, few millions will die in the process. Only about a hundred will reach the egg, and only one sperm will enter the egg to fertilize it. What happens to the other sperm; once the one sperm has fertilized the egg, is unknown, however, it possible that they will live for three-four days before they die.

What are Chromosomes?
Chromosomes are organelles of a cell which contain chemical structures called genes. These genes are the structure of DNA; which has a full set of instructions to create different parts of a human baby, for example, such instructions can be to give the baby brown eyes, blue eyes, blond hair, green eyes, be tall, be short, and have a naturally large BMI or small BMI etc. In sexual reproductions, when meiotic cells replicate themselves; each cell will receive a full set of genes; with the same instruction to create a particular type of baby. Once the cell has been instructed to become a part of the baby's body, it cannot change; and will only function for that area of body which it's been instructed to.

Hereditary
Alleles
Alleles are genes which control the physical characteristics of a newborn baby. The alleles are separated into two categories, therefore, the two categories are:
- **Dominant Allele:** Alleles that are Dominant are shown as (B).
- **Recessive Allele:** Alleles that are Recessive are shown as (b).

Example of Alleles affecting Physical Characteristics
When a male and female cat mates, the fathers sperm will travel to the uterus; carrying the dominant gene for long hair; called (B). The mother cat's egg will have recessive gene for short hair called (b). When the sperm fuses with the egg, in the process of fertilization to create a zygote, the dominant gene will take over the recessive gene. This means that the new gene will be (Bb) and the kitten's fur will be long. This example not only goes with fur, but also goes with every physical characteristic of any animal body whether it may be hair, length of hair, bones structure, shape of eyes, colour of eyes, shape of nose, and many more. For example, if the dominant gene of the father is green eyes, and the mother recessive gene is for blue eyes; the dominant gene will take over, making the new born baby have greener colour in its eyes than blue. Just because the babies eyes are greener does not mean the recessive gene

has been fully taken over since small colours of blue can still be seen but not as clearly as the green colour of the eyes.

Genotypes and Phenotypes
Although a brother and sister may have brown eyes, the genes which give them their brown colour can be the same or completely different. This can be determinate by (Bb) and (BB). Genotype explains how two different people, who are connected by blood and have the same phenotypes, still have different genes. As for phenotypes, it explains how two siblings have the same type of gene.

Meiosis and Mitosis

What is Meiosis?
Meiosis is a cell division process; where the cells divide and create gametes; therefore, these gametes are eggs, ovaries and sperm. The cell division of meiotic cells create four gamete cells, which are sperm in males, while in females four gamete cells will produce one mature egg; which is ready for fertilization.

What is Mitosis?
Mitosis is a cell division process where chromosomes will combine and separate, in this process of replication, in order to create more cells. Chromosomes are thread like structures, made from chromatids; which hold the genes needed for the human being to be created and born. This separation creates two daughter cells; with a diploid number of chromosomes. In a human being; the mitosis cells will have 46 chromosomes in their cells. This separation will continue with each cell having the diploid number of chromosomes.

Meiosis Procedure	Mitosis Procedure
1. **Interphase:** The chromosomes are not visible in the nucleus membrane of the cell. This is due to the chromosomes not being coiled.	1. **Prophase1:** The homologous chromosomes, centrioles, and organelles are visible in the cell. The chromosomes at this point are identical.
2. **Prophase:** The coiled chromosomes, centrioles, and organelles are present in the cell.	2. **Metaphase1:** Homologous chromosomes begin to line alongside each other in the middle of the cell.
3. **Metaphase:** The chromosomes line up in the middle of the cell and start to shorten as well as thicken. Centrioles in the cells will also connect to the chromatids of the chromosomes.	3. **Anaphase1:** The nuclear membrane disappears, which begins the cell division. In this process, the chromosomes will move to opposite end of the cells and create two cells; with haploid number of chromosomes.
4. **Anaphase:** The nuclear membrane in the cell disappears and the chromatids start to pull apart from each other to the opposite ends of the cells.	4. **Telophase1:** Cells with haploid number of chromosomes. These chromosomes have now become two chromatids.
5. **Telophase:** A new nuclear membrane formed around each set of the two new cells; therefore, this causes the cell to start dividing. When cell division is complete; two daughter cells are made, each being identical, and having the diploid number of chromosomes in their structure.	5. **Anaphase2:** Second cell division takes place which separates the chromatids into four chromatids.
	6. **Telophase2:** Finally, four gametes are formed. These gametes will rather become four separate sperms; or one mature egg.

What is the Difference between Mitosis and Meiosis Cells
- **Mitosis Cells:** Mitosis cells are known as somatic cells. These types of cells take place and replicate only in animals and plants for growth and replacement. Examples of these somatic cells are blood cells, epidermal cells, and epithelial cells. Mitosis cells are not sexual cells of any kind, however, the exception to this; is when mitosis cells replicate and covert themselves into sex/gamete cells. Even though mitosis cells can covert to meiosis cells, this still makes the sex cells meiotic and not meiotic. Mitosis cells contain the diploid number of chromosomes; which is 46.

- **Meiosis Cells:** Meiosis cells are cells which convert to sex/gamete cells, therefore, they replicate to create sperm and eggs, which are important for the reproduction of a baby. These cells only contain the haploid number of chromosomes. This is why sperm and eggs only contain 23 chromosomes in its structure. Gamete/sex cells will only attain the diploid number of chromosomes, where they both fuse in the process called fertilization.

Summary of Differences between Meiotic and Meiotic Cells

Mitotic Cells	Meiotic Cells
Replicate somatic cells.	Replicate Gamete/sex cells.
The diploid number of chromosomes are pass into the two daughter cells in cell division	The haploid number the chromosomes are passed onto the gamete cells in cell division.
The two daughter cells are twins and are completely identical in their chromosomes and genes	The homologues cells in meiotic cells are mixed up and are different from each other.
Mitotic cells only occur in asexual reproduction, therefore, they are able to create clones of their parents. Example of clones that came from their parents is plants.	Meiotic cells occur in sexual reproduction and will have variation between the child and their parents.

Development of the Embryo
The development of the embryo followed by the embryos preparations before labour is listed in days and weeks below.

First Stages of Development
- **Week 1-25 Days**: From week one, fertilization takes place in the womb. The fertilized egg will already be determined if it is a boy or a girl; burrow itself into the placenta and start cell division. After 7-10 days, would have grown into 256cells. The baby, in the egg which has converted to a zygote, will be 1/100 of an inch long and will have formed a tiny heart which beats, also. By day 20; the nervous system of the baby would have been created.

- **Four-Five Weeks:** After the 25th day, the circulatory system, blood circulation, will have been created in the baby; as well as the liver, lungs, larynx, inner ears, pancreas, and stomach. Around week five, the baby will measure 7-9mm from its head to its bottom. Its kidneys will start to be created.

- **42 days - Week6:** The skeletal system of the baby is created, and brain develops as well, brain waves of the baby should be detected at this time. Around week 6 all organ of the adult body should be created, and the hardening of bones, called ossification, begins.

- **Week 7- Week 8:** At week 7, the baby estimate length is 13-17 mm long and weighs about a gram. At week 8, the baby will weigh 4 grams and have the length of 27-35mm; also, the upper lip, external ears, and external genitals should be created and visible to the human eye in an ultrasound scans.

- **Week 9- Week 10:** When week 9 of the development arrives, the baby should weight 7 grams, and the iris of the eyes and fingers nails are visible. Around week 10, the baby should be able to swallow, move its tongue, using its muscles; and be able to make a fist. The baby's brain will also take structure and function with all the other systems of the body.

- **Week 11- Week 12:** The development of the baby's teeth and villi have formed, the pancreas will be secreting insulin. Around week 12, the baby will practice breathing; as well as be producing urine which will go out through the placenta.

Placenta
The placenta is a lining which is created from balls of cells, therefore, the placenta prepares itself for fertilization; and the egg to implant itself into it, however, if no fertilization occurs, the placenta will come out of the vagina during the menstrual cycle. If the egg is fertilized and

implants itself to the uterus; the umbilical cord of the baby will be connected to the placenta. Therefore, after the babies cardiovascular system is created and working; the placenta role is to pass oxygen, glucose, amino acids, and salts to the baby in its blood; while carbon dioxide, urea and waste are taken out by the living and growing tissue. Other roles of the placenta; is to secrete the hormones progesterone and oestrogen, to keep the pregnancy going and healthy; as well as prepare the mother mammary glands for breast feeding.

Second Stages of Development
- **Week 14-Week15:** Mothers will feel the movement of their baby in the womb around the 14th week. Around week 15, the baby should weight an estimate of 142 grams.

- **Week 16- Week 17:** The baby will start to grow eyelashes, and respond to touch, movements and sounds. If the baby is a girl, it will start to form primitive eggs in their ovaries and the permanent bud, which will grow permanent teeth, are formed behind the babies soon to be milk teeth.

- **Week 19 – Week 20:** The baby's head will move to a head down position; which points towards the uterus and out of the vagina. The baby will should weight an estimate of 359 grams; as well as develop eyelashes and grow larger.

- **Week 21- Week 24:** During these weeks, the baby will grow larger and larger.

Final Stages of Development
- **Week 27- Week 28:** Movements the baby shall be noticed and it shall weight an estimate of three pounds in the womb.

- **Week 29 – Week 38:** The eyes of the baby will dilate from the light changes in the womb; as its head turns towards the uterus of the woman. The baby shall put an estimate of 5 pounds on every week, and the final developments and refinements will be made before labour begins. Once the baby is out of the uterus, it may be the length of 20 inches.

Birth/Labour
Labour begins when the fetus of the babies head faces toward the uterus; with the head above the cervix. After this happens, muscles in the uterus start to contract rhythmically; therefore, these contractions will become frequent and much stronger. The cervix walls will start to widen, and open slowly; to let the babies head come through the cervix. Also, the baby will be moved out of the womb by the contractions in the abdominals. Sometime in the process of labour, the water sacks will break; and water will seep out through the vagina, therefore, this is the first sign of labour taking place. After a few more strong and frequent contractions, in which the woman in labour will be in pain; the baby will come head first out of the vagina; through the uterus. The baby will be born with its umbilical cord still attached to the bellybutton, therefore, after the birth; it will be tied and cut. The umbilical cord attached to the baby; will shrivel away after a few days. After the birth, the placenta; will break away from the uterus and come out of the vagina as afterbirth. The baby will start to cry too; since the temperature of its surrounding will drop; which will cause it to take its first breath and cry.

Induced Birth
If the baby does not give signs of labour after 38 weeks; doctors will induce the baby to come out by breaking the membrane of the amniotic sac. Another way to induce labour is to inject a hormone called Oxytocin; which triggers the muscles of the uterus to contract and start labour.

Sexes of the Baby
The sexes of a baby are determined by the chromosome structure. The picture below shows the male and female chromosomes.

Picture:

(Taken from http://open.jorum.ac.uk)
Summary of Sex Determination
- XX = Female
- XY = Male

END

References
Digestive System
(Natural-Reflux-Core.com, 2006-2011, Stomach acid is there for a reason. What is this reason and how much are we supposed to have?, http://www.natural-reflux-cure.com/stomach-acid.html, Sunday 15th January 2012)
(Natural-Reflux-Core.com, 2006-2011, With all this acid, why doesn't the stomach digest itself?, http://www.natural-reflux-cure.com/digest-itself.html, Sunday 15th January 2012)
www.Buzzle.com,2000,2011 and 2012, List of Digestive Enzymes, http://www.buzzle.com/articles/list-of-digestive-enzymes.html, Monday 12th March 2012)
(Infoplease.com, 2000-2011, Vitamins and Minerals, http://www.infoplease.com/cig/biology/vitamins-minerals.html, Friday 9th December 2011)
Respiratory System
(jcc.net, 2002, Cellular Respiration Overview, http://staff.jccc.net/pdecell/cellresp/respintro.html, Thursday 29th March 2012)
(livestrong.com, 2012, How does the Cardiovascular system work with the Respiratory System?, http://www.livestrong.com/article/18606-cardiovascular-system-work-respiratory-system/, Thursday 29th March 2012)
(wisteme.com, 2010, How does Cellular Respiration work?, http://www.wisteme.com/question.view?targetAction=viewQuestionTab&id=19338, Thursday 29th March 2012)
Circulatory System
(Answers.com, 2012, If Blood is made inside your bone marrow, how does it get out?, http://wiki.answers.com/Q/If_blood_is_made_inside_your_bones_how_does_it_get_out, Wednesday 28th March 2012)
(goodtoknow.co.uk,2012 The bone marrow, stem cells, and blood cell production, http://www.goodtoknow.co.uk/wellbeing/134635/The-bone-marrow--stem-cells-and-blood-cell-production, Wednesday 28th March 2012)
Renal System
(Yahooanswers.com, 2012, Outline the role of the Loop of Henle, http://answers.yahoo.com/question/index?qid=20090303171702AADuc3k, Wednesday 28th March 2012)

(youtube.com, 2012, Uniary system, the nepheron, http://www.youtube.com/watch?v=aQZaNXNroVY, Wednesday 28th March 2012)
Immune System
(NHS Choices, 2010, Atheletes foot Causes, http://www.nhs.uk/Conditions/Athletes-foot/Pages/Causes.aspx, Saturday 14th January 2012)
(NHS Choices, 2010, Vaginal Thrush, Vaginal Thrush causes, http://www.nhs.uk/conditions/thrush/Pages/Introduction.aspx, http://www.nhs.uk/Conditions/Thrush/Pages/Causes.aspx, Saturday 14th January 2012)
(EMedicine Health, 2012, Yeast infection skin rash, http://www.emedicinehealth.com/yeast_infection_skin_rash/page2_em.htm#Yeast%20Infection%20Skin%20Rash%20Causes, Saturday 14th January 2012)
(History of the Unvierse, 2011, Bacterial Benefits, http://www.historyoftheuniverse.com/bactbene.html, Saturday 14th January 2012)
(solopugid.com, 2012, Protoctisits, http://www.solpugid.com/cabiota/Protoctista.htm, Saturday 14th January 2012)
(bioumass.edu, 2012, Protoctista, http://www.bio.umass.edu/biology/conn.river/protoc.html, Saturday 14th January 2012)
(Science of Learning, 2007-2012, The Bodies First Line of Defence, http://www.sciencelearn.org.nz/Contexts/Fighting-Infection/Science-Ideas-and-Concepts/The-body-s-first-line-of-defence, Sunday 15th January 2012)
(Natural-Reflux-Core.com, 2006-2011, Stomach acid is there for a reason. What is this reason and how much are we supposed to have?, http://www.natural-reflux-cure.com/stomach-acid.html, Sunday 15th January 2012)
(Natural-Reflux-Core.com, 2006-2011, With all this acid, why doesn't the stomach digest itself?, http://www.natural-reflux-cure.com/digest-itself.html, Sunday 15th January 2012)
(http://www.niaid.nih.gov, 2012, T cells, http://www.niaid.nih.gov/topics/immuneSystem/immuneCells/Pages/tcells.aspx, Sunday 15th January 2012)
(http://www.niaid.nih.gov, 2012, B cells, http://www.niaid.nih.gov/topics/immuneSystem/immuneCells/Pages/bcells.aspx, Sunday 15th January 2012)
(http://www.niaid.nih.gov, 2012, Phagocytes and their relatives, http://www.niaid.nih.gov/topics/immuneSystem/immuneCells/Pages/phagocytes.aspx, Sunday 15th January 2012)
(Wisegeekcom, 2003-2012, What is Phagocytoses?, http://www.wisegeek.com/what-is-phagocytosis.htm, Sunday 15th January 2012)
(Human Anatomy, 2006, Animation-Phagocytosis, http://highered.mcgraw-hill.com/sites/0072495855/student_view0/chapter2/animation__phagocytosis.html, Sunday 15th January 2012)
(Cleveland Clinic, 2012, Inmmaflation, http://my.clevelandclinic.org/symptoms/inflammation/hic_inflammation_what_you_need_to_know.aspx, Monday 16th January 2012)

(Sports Injury Clinic, 2012, Inflammation and Tissue Healing, http://www.sportsinjuryclinic.net/sport-injuries/general/inflammation, Monday 16th January 2012)

Renal System

(Yahooanswers.com, 2012, Outline the role of the Loop of Henle, http://answers.yahoo.com/question/index?qid=20090303171702AADuc3k, Wednesday 28th March 2012)

(youtube.com, 2012, Uniary system, the nepheron, http://www.youtube.com/watch?v=aQZaNXNroVY, Wednesday 28th March 2012)

Books

(Bradfield and Potter, 2009, Edexcel IGCSE Biology, Edinburgh Gate, Harlow, Essex, CM20 2JE, Pearson Education Limited.)

(Mackean, 1996, GCSE Biology second edition, 50 Albemarle Street London W1X 4BD, John Murray)

Endocrine System

(CrystalLinks.com, 2012, Third Eye-Pineal Gland, http://www.crystalinks.com/thirdeyepineal.html, Thursday 19th April 2012)

(Answers.com, 2012, What are the functions of the Pineal Gland?, http://www.answers.com/topic/what-are-the-functions-of-the-pineal-gland, Thursday 19th April 2012)

(Answers.com, 2012, What is the function of the Pineal Gland?, http://wiki.answers.com/Q/What_is_the_function_of_the_pineal_gland&altQ=What_is_the_pineal's_function, Thursday 19th April 2012)

(Answers.com, 2012, Does Melatonin really help you sleep?, http://wiki.answers.com/Q/Does_Melatonin_really_help_you_to_sleep, Thursday 19th April 2012)

(About.com/Biology, 2012, Pineal Gland, http://biology.about.com/od/anatomy/p/pineal-gland.htm, Thursday 19th April 2012)

(yourhormones.info, 2011-2013, Pituitary Gland, http://www.yourhormones.info/glands/pituitary_gland.aspx, Thursday 19th April 2012)

(biology.clc.uc.edu, 1996-2004, Endocrine System, http://biology.clc.uc.edu/courses/bio105/endocrin.htm, Thursday 19th April 2012)

(Pituitray.org.uk, 2012, Pituitary Overview, http://www.pituitary.org.uk/content/view/19/28/, Thursday 19th April 2012)

(Mcmillan.org.uk, 2012, Pituitary Gland, http://www.macmillan.org.uk/Cancerinformation/Cancertypes/Brain/Typesofbraintumours/Pituitarytumours.aspx#DynamicJumpMenuManager_6_Anchor_2, Thursday 19th April 2012)

(About.com/Biology, 2012, Hypothalamus, http://biology.about.com/od/anatomy/p/Hypothalamus.htm,Thursday 19th April 2012)

(youandyourhormones.info, 2011-2013, Hypothalamus, http://www.yourhormones.info/glands/hypothalamus.aspx, Friday 20th April 2012)

(Ivy-rose.co.uk, 2012, Human Adrenal Glands, http://www.ivy-rose.co.uk/HumanBody/Endocrine/Adrenal_Glands.php, Friday 20th April 2012)

(bookrags.com, 2012, Adrenal Glands and Hormones, http://www.bookrags.com/research/adrenal-glands-and-hormones-wap/, Friday 20th April 2012)

(endfatigue.com, 2011, The Adrenal Gland, http://www.endfatigue.com/health_articles_a-b/Adrenal-the_adrenal_gland.html, Friday 20th April 2012)

(Paitent.co.uk, 2012, The Thyroid and Parathyroid Glands, http://www.patient.co.uk/health/Thyroid-and-Parathyroid-Glands.htm, Saturday 21st April 2012)

(Parathyroid.com, 1996-2012, Parathyroid Glands, http://www.parathyroid.com/parathyroid.htm, Saturday 21st April 2012)

(Wisegeek.com, 2003-2012, What is the Thyroid Gland?, http://www.wisegeek.com/what-is-the-thyroid-gland.htm, Saturday 21st April 2012)

(innerbody.com, 1999-2012, Thymus, http://www.innerbody.com/image_endoov/lymp04-new.html, Saturday 21st April 2012)

(erikanderson.net, 2012, Thymus, http://www.erikanderson.net/endocrine/thymus.shtml, Saturday 21st April 2012)

(howstuffoworks.com, 1998-2012, Thymus, http://science.howstuffworks.com/environmental/life/human-biology/immune-system6.htm, Saturday 21st April 2012)

(Maricopa.edu, 1992- 2010, The endocrine System, http://www.emc.maricopa.edu/faculty/farabee/biobk/biobookendocr.html, Saturday 21st April 2012)

(Bupa, 2011, Low Blood Pressure (Hypotension), http://www.bupa.co.uk/individuals/health-information/directory/l/low-blood-pressure#textBlock202714, Monday 21st November 2011)

(Yahooanswers.com, 2012, What's the Difference Between an endocrine and exocrine gland?, http://uk.answers.yahoo.com/question/index?qid=20100124201353AAJIEr3m, Saturday 21st April 2012)

(Yahooanswers.com, 2012, What are Endocrine and Exocrine Glands?, http://answers.yahoo.com/question/index?qid=20090716193506AAdZGyf, Saturday 21st April 2012)

(hormone.org, 2012, Endocrine Glands, http://www.hormone.org/Endo101/page2.cfm, Saturday 21st April 2012)

Nervous System

(About.com, 2012, Peripheral Nervous system, http://biology.about.com/od/organsystems/a/aa061804a.htm, Monday 23rd April 2012)

(yahooanswer.com, 2012, What's the difference between involuntary and voluntary muscles?, http://answers.yahoo.com/question/index?qid=20080306111132AAvoXJn, Monday 23rd April 2012)

(weisegeek.com, 2003-2012, What is a Interneuron?, http://www.wisegeek.com/what-is-an-interneuron.htm, Monday 23rd April 2012)

(Answers.com2012, What is an axon terminal?, http://wiki.answers.com/Q/What_is_an_axon_terminal, Friday 27th April 2012)

(Yahooanswers.com, 2012, What is the Node of Ranviers Function?, http://answers.yahoo.com/question/index?qid=20110815125838AAJsi9V, Friday 27th April 2012)

(biology-online.org, 2008, Schwann Cells, http://www.biology-online.org/dictionary/Schwann_cells, Friday 27th April 2012)

(blustein.tripod.com,2012, Schwann Cells, http://blustein.tripod.com/Schwann_Cells/schwann_cells.htm, Friday 27th April 2012)

(Yahooanswers.com, 2012, Similarities between nervous and endocrine system?, http://answers.yahoo.com/question/index?qid=20080929011247AA63ji7, Wednesday 25th April 2012)

(Yahooanswers.com, 2012, What are the Similarities and Differences Between the Nervous System and Hormonal System?, http://uk.answers.yahoo.com/question/index?qid=20090211075221AAWdP1T, **Wednesday 25th April 2012)**

(Yahooanswers.com, 2012, Differences between Nervous and Endocrine System?, http://answers.yahoo.com/question/index?qid=20070503041800AAQtgfd, **Wednesday 25th April 2012)**

(Answers.com, 2012, What are the similarities between the nervous and hormonal system?, http://wiki.answers.com/Q/What_are_the_similarities_between_the_nervous_system_and_the_hormonal_system, **Wednesday 25th April 2012)**

(Answers.com, 2012, what is Negative Feedback Mechanism, http://wiki.answers.com/Q/What_is_a_negative_feedback_mechanism, **Wednesday 25th April 2012)**

(blurtit.com, 2012, What is the difference Between the Nervous System and the Endocrine System?, http://www.blurtit.com/q495666.html, **Wednesday 25th April 2012)**

Reproductive System

How do Homes Control the Menstrual Cycle and Pregnancy?

(netdocter.co.uk, 1998-2011, Female Hormones, http://www.netdoctor.co.uk/womenshealth/features/hormone.htm, **Saturday 21st April 2012)**

(Yahooanswers.com, 2012, How do hormones control the menstrual cycle?, http://uk.answers.yahoo.com/question/index?qid=20090923120414AA0rmFt, **Saturday 21st April 2012)**

(netdocter.co.uk, 1998-2011, The Menstrual Cycle, http://www.netdoctor.co.uk/health_advice/facts/menstruation_cycle.htm, **Saturday 21st April 2012)**

(biotopics.co.uk, 2012, Hormonal Control of the Menstrual Cycle, http://www.biotopics.co.uk/newgcse/menstrualhormones.html, **Saturday 21st April 2012)**

(womenshealthgrow.gov, 2009, Menstruation and the Menstrual Cycle, http://www.womenshealth.gov/publications/our-publications/fact-sheet/menstruation.cfm, **Saturday 21st April 2012)**

(medicinenet.com, 1996-2012, Menstrual Cramps, http://www.medicinenet.com/menstrual_cramps/page2.htm, **Thursday 26th April 2012)**

(Yahooanswers.com, 2012, what are the side affects that come with a period?, http://answers.yahoo.com/question/index?qid=20080115152420AACHFIQ, **Thursday 26th April 2012)**

(http://peacepigeon.tripod.com, 2012, Time Line of the Development of the Human Fetus, **Monday 7th May 2012)**

Books

(Bradfield and Potter, 2009, Edexcel IGCSE Biology, Edinburgh Gate, Harlow, Essex, CM20 2JE, Pearson Education Limited.)

(Mackean, 1996, GCSE Biology second edition, 50 Albemarle Street London W1X 4BD, John Murray)